XXXIX

PENNSYLVANIA IRON MANUFACTURE

IN THE EIGHTEENTH CENTURY

PENNSYLVANIA

IRON MANUFACTURE

IN THE

EIGHTEENTH CENTURY

BY

ARTHUR C. BINING

[1938]

REPRINTS OF ECONOMIC CLASSICS

AUGUSTUS M. KELLEY · PUBLISHERS

NEW YORK 1970

(Harrisburg: Pennsylvania Historical Commission, 1938)

REPRINTED 1970 BY
AUGUSTUS M. KELLEY · PUBLISHERS
NEW YORK NEW YORK 10001

ISBN 0 678 00678 4
LCN 72 120547

PRINTED IN THE UNITED STATES OF AMERICA
by SENTRY PRESS, NEW YORK, N. Y. 10019

TO B. F. FACKENTHAL, JR., Sc. D., LL.D.
Outstanding Student and Friend of Pennsylvania History,
and President of the Bucks County
Historical Society

Contents

Page

Preface 7

I. First Attempts at Ironmaking in America....... 11

II. The Iron Plantations......................... 29

III. The Establishment of the Industry............. 49

IV. The Technique of Iron Manufacture............ 67

V. Improvements and Inventions.................. 95

VI. The Workers 107

VII. The Ironmasters 131

VIII. Relations with England...................... 149

IX. The Progress of the Iron Industry............. 169

Appendices 187

Index 215

Illustrations

Page

Rutter Mansion, Pine Forge.................Frontispiece

Typical Cold-Blast Charcoal Furnace................. 15

Refinery Forge 18

Hay Creek Forge.................................... 21

Mansion House, Warwick Furnace................... 30

Mansion House and Store, Sally Ann Furnace........ 32

Old Buildings, Warwick Furnace.................... 33

Old Office Building, Warwick Furnace............... 37

Mansion House, Durham Iron Works................. 42

Ruins of Alliance (Jacob's Creek) Furnace........... 44

Ruins of Hopewell Furnace........................ 50

Ruins of Sally Ann Furnace........................ 52

Plan of Durham Iron Works Property............... 54

Recent View of Cornwall Mines.................... 58

Ruins of Pine Grove Furnace...................... 59

Bloomery Forge 63

Charcoal House, Elizabeth Furnace................. 72

Charcoal Pile in Process of Being Charred........... 74

Plan of Eighteenth Century Cold-Blast Charcoal Furnace 78

Stove Plate 81

Hay Creek Forge, Front View...................... 83

Illustrations

	Page
Blister Steel Furnace	86
Forging an Ancony at Refinery Forge	87
Plan of Rolling and Slitting Mill	89
Stove Plate	96
Franklin Open Fireplace	98
Fireplace Cast at Hopewell Furnace	100
Ten Plate Stove	102
Hammer Heads and Wedge	113
Refinery Forge Hammer	116
Stove Plate Cast at Durham Iron Works	120
Tilt Hammer, Hay Creek Forge	122
Six Plate Draft Stove	135
Daniel Udree, Ironmaster, Oley Furnace	138
George Taylor	143
Plating Mill	146
James Logan	151
Rolling and Slitting Mill	156
Oak Pinion and Gears, Hay Creek Forge	163
Mansion House, Pine Grove Furnace	172
James Logan Fireback	176
Stove Plate Cast at Durham Iron Works	181

PREFACE

THIS monograph is the result of intensive research on one unit of eighteenth century American industrial life. The chief purpose of this study is to present in some detail an account of the origin and progress of the Pennsylvania iron industry during its first century of development, especially its social and economic aspects. The story of the pioneers who laid the foundations for Pennsylvania's later great iron and steel industry, the details of the early charcoal-iron industry including its processes, output and markets, the close association between ironmaking and agriculture, the many problems that had to be solved as a result of establishing an industry under new conditions in a new country, and the place and importance of iron manufacture in an agrarian civilization, present a most interesting and even fascinating chapter in our history.

The iron industry in the American colonies had its real beginnings in New England in the seventeenth century, but it was not until the first part of the eighteenth century that it began to show rapid growth. It was in this period that Pennsylvania started on her industrial career. After the middle of the eighteenth century, Pennsylvania became the foremost iron producing center in the colonies. From this time on, steady growth was made, which prepared the way for the rapid expansion and the many changes that were to follow.

The chronological limits of this work have been set from the period of the origin of iron manufacture in Pennsylvania to the year 1800. In order to present a true perspective the background of the American iron industry has been briefly sketched. The date for concluding the study has been chosen partly for convenience and partly because changes were beginning to take place that finally brought in a new industrial

7

era. After 1800, the application of large quantities of capital to industry, the improvements made in transportation, the gradual broadening of markets due to westward expansion, and the beginning of technological changes, were to result in giving industrial life a different complexion. The charcoal-iron industry in many of its branches remained throughout the nineteenth century and even later, but during this period new processes and forms of organization were being established that marked the doom of the earlier methods of iron production.

Iron manufacture in all parts of the country during the eighteenth century was quite similar to that of Pennsylvania. The forms of organization, the types of ironworks, the methods of production, the means of transportation, and the problems concerning markets were quite similar. Only in details were there differences and these were few. The broad outlines of this study, therefore, may be applied to iron manufacture in any part of America where iron was produced in the eighteenth century.

The illustrations used in this volume have been obtained from a variety of sources. Several have been furnished by Dr. B. F. Fackenthal, Jr., Mr. Richard Peters, Jr., Mr. Charles B. Montgomery, and by the Bucks County Historical Society. Many are from the author's collection. As no contemporary pictures of eighteenth century American ironworks in operation could be found, a few illustrations of European ironworks of the period are included. From written descriptions it is evident that the American works were similar in all details to those of Europe after which they were patterned. These have been taken from D. Diderot and J. B. Alembert's *Recueil de Planches sur les Sciences, les Arts Libéraux, et les Arts Méchaniques* (Paris, 1765, Livourne, 1774). The contemporary plan of an American rolling and slitting mill is from the *Transactions* of the American Society of Mechanical Engineers, 1881. These have been included to make clearer the descriptions of eighteenth century methods and processes. Of the many old mansion houses that still remain, as well as many old ruined furnace stacks and other evidences of an almost-forgotten industry, only a few have been chosen for illustrative purposes. They have been selected as being representative.

The author desires to express his appreciation for the courtesies extended and the help given by the library staffs of the Historical Society of Pennsylvania, the American Philosophical Society, the Historical Society of Western Pennsylvania, the Pennsylvania State Library, the Carnegie Library of Pittsburgh, the many historical societies of Pennsylvania, the Library of Congress, the Baker Library of the Business Historical Society, the New York Public Library, the William L. Clements Library of American History, and to the officials of the Public Record Office, London. He acknowledges gratefully his obligations to many who have given constructive criticisms, suggestions and encouragement, including Dr. B. F. Fackenthal, Jr., of Riegelsville, Mr. Richard Peters, Jr., of Philadelphia, Mr. Charles B. Montgomery of Philadelphia, Professor Asa E. Martin of Pennsylvania State College, Professor John W. Oliver of the University of Pittsburgh, Professor Lawrence H. Gipson of Lehigh University, Professor Charles M. Andrews of Yale University, the late Professor Albert E. McKinley, and also Professor St. George L. Sioussat, Professor Roy F. Nichols, Professor William E. Lingelbach, Professor Emeritus Edward P. Cheyney of the University of Pennsylvania, and to Miss M. Atherton Leach of Philadelphia. The author takes this opportunity of thanking also the Faculty Research Committee of the University of Pennsylvania and the Social Science Research Council for their interest and for grants-in-aid to carry out the project. Grateful acknowledgment is offered to the Pennsylvania Historical Commission, Major Frank W. Melvin, Chairman, which has made possible the publication of this work, and to Mr. Donald A. Cadzow, Anthropologist of the Pennsylvania Historical Commission, who has seen it through the press to completion. To many others who have assisted the author tenders thanks.

UNIVERSITY OF PENNSYLVANIA   ARTHUR CECIL BINING
MAY, 1938

Rutter Mansion, Pine Forge

CHAPTER I

FIRST ATTEMPTS AT IRONMAKING IN AMERICA

IT IS not probable that the natives of America worked iron in any way for practical uses before the white man came.[1] Certain tribes on different parts of the continent used the reddish oxide as pigment for their war paints,[2] frequently meteors fallen from the sky were beaten into objects of worship, and occasionally small, loose pieces of hematite were made into symbols of good luck.[3] But the scattered masses of iron ores remained, as in ages past, untouched. When Christopher Columbus discovered the West Indies in 1492, he found the islands inhabited by peaceful natives, who were not antagonistic to the adventurous Spaniards. He was quite sure that these people of San Salvador, the island he first reached, had no knowledge of the use of iron or of edged weapons of any sort, for they showed their ignorance of the latter by grasping the swords of the Spaniards, some even cutting themselves. Their stage of culture was a lowly one. They pointed their javelins with fishbones, and their darts with teeth of fish. They brought to the explorers such things as parrots, cotton thread in balls, and darts, for which they received in exchange red caps, small bells, strings of glass beads and other trinkets.[4]

While seeking feverishly for gold on many islands of the West Indies as well as for a route or passage to the East Indies, Columbus found no natives acquainted with the use of "iron, steel or firearms."[5] The fourth voyage, begun in 1502, was a pathetic attempt to discover the mainland of Asia in order to remove the ignominy which had fallen upon him. In the course of this last voyage, the great explorer came in contact with the Indians from the mainland of America. Among the articles in which they traded were mirrors and bracelets of gold, axeheads of copper, and flint knives. None of them, however, had any tools, utensils, or articles made of iron.[6]

When the Spaniards under Cortez reached Mexico in 1519, they found a comparatively well-developed civilization among the Aztecs. The accounts of early explorers, painted in glowing colors, have been found to contain much truth, as the result of the numerous remains which have been examined in more recent years. While agriculture was the main occupation of the Aztecs, manufactures and trade were developed to a relatively considerable extent. There was some specialization in industry, and regions became noted for certain products. Markets under government supervision were held regularly in specified places. Silver, lead and tin were mined at Tasco, and copper was taken from the mountains of Zacotollan. Tools made from an alloy of tin and copper, with the aid of silicious dust, could cut the hardest substances. Knives, razors and swords were made of "itztli" (obsidian—a dark, transparent, volcanic glass). Gold, found near the surface of the earth or in the beds of rivers, was cast into bars. Vessels of gold and silver were made and often carved with great skill. But the use of iron was unknown to the Aztecs in spite of the fact that ferruginous masses were scattered over the surface of their tableland.[7]

The Incas of Peru, likewise, had made much progress in the arts of civilization by the time Pizarro and his men discovered their country in 1527. Like the Aztecs of Mexico, they smelted and worked gold and silver.[8] The royal buildings and the tombs of the kings have been found to contain valuable vases and ornaments of elaborate workmanship. Their tools were of stone, gem, copper, and of an alloy of tin and copper. With such tools they cut even emeralds and other precious stones. But they did not use iron. Because of the lack of iron horseshoes, Pizarro on several of his marches compelled the Indian smiths to shoe the horses of his troops with silver, which apparently proved satisfactory.[9] Even though the soil of Peru contained iron, the natives did not employ it in any way.[10]

The Indians of the north were far inferior in civilization to those of Mexico and Peru. Of their prehistoric remains, stone objects are the most common. There have been found in large quantities stone weapons, as arrow-heads, knives, axes and celts; stone tools of many kinds, such as hammers, gouges, scrapers, drills and chisels; utensils which include soapstone vessels, mortars and pestles; and ornaments of stone, as pendants, beads and disks.[11] All iron objects unearthed in the old Indian mounds, except a few of meteoric iron, are of European manufacture or date after the time when Europeans first made contact with the Indians.[12]

The civilization of the Pueblo region of the West was not highly developed. The ruins of their "apartment houses"—the strange homes of these Indians—have yielded numerous objects of stone, bone and wood. In addition, cotton cloth, mats and baskets of osier, cords of yucca fibers, and pottery of various types, remain as evidence of a race that has passed away. No traces of iron implements or tools, however, have ever been discovered in the remnants and debris which still exist.[13]

It may seem strange that the Indians did not use iron, the most plentiful metal found on the continent. Among the more primitive Indians of the North, iron manufacture might not be expected. But among the highly skilled artisans of Mexico and Peru, the question as to why they did not work iron appears to be a fair one. The answer may be found in the fact that gold, silver, copper and lead are often found in a "free state"—almost pure, while iron is never obtained free from combinations with oxygen, sulphur, phosphorus, silica and other materials. To isolate it from these components is a relatively difficult task. Other nations of antiquity found the means of working iron because they lacked the precious metals of Mexico and Peru, whereas in these countries there was no need to work the refractory iron when abounding deposits of gold, silver and copper could be used for their requirements.

Under the rule of the Spaniards, the mining of precious metals continued and developed to a great extent.[14] Of the thousands of mines which were explored and exploited by the Spaniards, many became famous. Indian and Negro workers produced great wealth for their masters. Large quantities of gold and silver were sent to Spain, much of which went to finance the wars of Europe. No attempt was made by the Spaniards however, to utilize the iron ores of the new Spanish Empire in America.[15]

During the first attempts made by the English to establish

colonies in America, iron was discovered. The expedition of Sir Richard Grenville, which was sent out by Sir Walter Raleigh in 1585 found iron ore in the region that many years later became North Carolina. Thomas Hariot, the historian of the expedition, wrote a short account of the new country.[16] In it he stated that iron ore was seen in many places. Since there was "an infinite store of wood" at hand, and because of the "want of wood and deereness thereof in England," he pointed out it would be profitable to manufacture iron in the new colony.[17] Famine and war with the Indians put an end to the attempts at settlement, and to Hariot's dream of iron manufacture. No iron was made in North Carolina until after the opening of the eighteenth century.[18]

About twenty years later the Virginia Company of London was chartered to colonize Virginia and within a short time the beginning of the first permanent English colony was made. Even before settlement the company planned to smelt the iron ores of the New World and expected that the iron thus produced would be used in England for the manufacture of iron articles. Among the ambitious plans of the company was the production of glass and other commodities. As early as 1608 a ship sailed away from Virginia carrying iron ore, which was smelted in England and sold to the East India Company.[19] During these early years, however, the visionary dreams of dotting the wilderness of a new continent with blazing iron furnaces for the benefit of the mother country were not immediately carried out, for after the early trying years it was discovered that tobacco could be profitably grown. The rich bottom lands were quickly exploited in growing tobacco for the European market and the industrial projects were neglected.

A group of men, however, with industrial leanings, known as the Southampton Adventurers, under patents granted by the Virginia Company of London decided to build ironworks. A part of the expected profits were to be used in the Christianizing and educating of Indian children. By 1621 the construction of the ironworks was well under way.[20] They were located on the west side of Falling Creek,[21] sixty-six miles above Jamestown. The lighting of the fires at the works was the signal for the terrible massacre by the Indians on March 22, 1622.[22] John Berkeley, who had charge of erect-

Typical Cold-Blast Charcoal Furnace

ing and conducting the works, and his workmen were killed, while the plant itself was completely destroyed. Before the Virginia Company had time to recover from this blow, its troubles with the English government increased. In 1624 its charter was revoked,[23] and Virginia was transformed into a royal colony. During the years that followed the catastrophe at the settlement just below the falls of the James River, suggestions and attempts were made from time to time to establish ironworks in Virginia, but without success for almost a hundred years, when Governor Spotswood began his enterprises.[24]

The colonial iron industry was first established in New England among the bog ores of the coastal region. In 1629, at the very beginning of the colonization of Massachusetts Bay, the Court of Assistants in London, commissioned an agent, Malbon, to make a survey of the new colony with the hope and object of organizing a company and erecting ironworks. A short time later, Thomas Graves, of Gravesend, Kent, received a similar commission.[25] Thus, as in Virginia, visionary industrial schemes were worked out. It was not until 1641, however, that John Winthrop, Jr., son of the governor, went to London to secure capital to begin the building of ironworks in Massachusetts Bay Colony. Two years later, Robert Bridges followed him, taking with him some specimens of bog ore from the ponds of the Saugus. Together they were successful, and the "Company of Undertakers for the Iron Works" was formed.[26] Not only was capital secured, but skilled workers were brought from England to produce iron. Winthrop had succeeded in interesting "eleven English gentlemen" in the project who subscribed £1,000 to establish the works. In 1644, the company was incorporated by the General Court, and obtained a monopoly for making iron in the colony for twenty-one years.[27] As the capital was insufficient, citizens were invited to subscribe to the project.[28] The company was freed from public charges and from taxation upon its stock. The proprietors and workers were exempted from military "trainings" and from "watchings" for the Indians that were required of other citizens.[29] Furnaces and forges were built at Lynn on the Saugus River, and Braintree,[30] about ten miles south of Boston.

The history of both works for fifty years was bound up

with litigation. Suits were brought by the company's agents against the company, workmen sought legal redress for wages long overdue, and creditors often obtained judgments. Lawsuits also arose because of damages to neighboring property caused by the breaking of dams.[31] A contemporary writer, amazed at the entanglement of litigation, stated that " . . . instead of drawing out bars for the country's use, there was hammered out nothing but contention and lawsuits, which was but a bad return for the undertakers." This statement was somewhat exaggerated, for quantities of excellent iron were smelted, cast and forged from the bog ores of Massachusetts at both these works.

It was not long, before the neighboring colony of Plymouth attempted to emulate Massachusetts Bay Colony in erecting ironworks. Timothy Hatherly, one of the founders of Scituate planned to manufacture iron and was granted lands near Mattakeeset Pond in 1650. The project did not mature and the lands reverted to the colony. It was not until 1702 that a blast furnace was built in this region.[32] However, by 1656 a bloomery forge for making bar iron directly from the ores, had been built in Plymouth colony by a company composed of the citizens of Taunton. The plant was known as the Raynham works. As was the case with other early ironworks, the hammers and the machinery were imported from England. During King Philip's War, the works withstood terrific Indian attacks. This forge was in operation for a period of about 200 years.[33]

Not long after the establishment of the Taunton forge in Plymouth the General Court of Massachusetts granted to the inhabitants of Concord and Lancaster permission to erect ironworks because, according to the charter, of the failure of the company at Lynn and Braintree to provide the colony with a sufficient amount of iron.[34] In 1674, Nathaniel and Thomas Leonard made a contract with John Ruck and others of Salem to forge iron at Rowley. This forge was not successful and its career soon came to an end.[35] A few other bloomery forges were established before the close of the seventeenth century in Massachusetts Bay Colony.

The manufacture of iron was begun quite early in the life of the colony of Connecticut. John Winthrop, Jr., enthusiastic over his industrial schemes, soon after organizing the colony at the mouth of the Connecticut River, petitioned the Massachusetts General Court for permission to lay out plantations and ironworks at Pequot (New London).[36] This was granted in 1644, but no evidence has been found to show that the works

Refinery Forge

were ever built.[37] In partnership with William Paine and others, however, Winthrop built works at New Haven. They consisted of a blast furnace and a forge which went into operation in 1657.[38] In an attempt to secure superior ores, vessels sailed from England with iron ore that was used at these works. Such ores were mixed with those of Connecticut. The plan proved to be too expensive and the local ores continued to be used. The early ironworks, built along the coast, had to rely chiefly on bog ores which were taken from bogs and ponds. It was not until the middle of the eighteenth century that the rich ore deposits of western Connecticut were discovered and utilized.[39]

The radical Puritan and eloquent preacher, Roger Williams, attempted to encourage the erection of ironworks in Rhode Island. In 1654, he asked John Winthrop, Jr., to assist in promoting an enterprise at Providence.[40] Soon afterwards, Joseph Jenks, a machinist and moulder who made the dies for the pine tree shillings and had worked out several inventions,[41] left Lynn for the purpose of building a bloomery at Pawtucket. This forge was successful for many years, but during the Wampanoag war with the Narragansett Indians in 1675, it was completely destroyed.[42] It is probable that other small forges were established in this colony before the century closed, but if so, no record has been left of them.

While several ironworks were established in New England before the seventeenth century closed, the dreams of many industrialists of an extensive iron industry in the New World had not come true. In 1673, Edward Randolph, commissioned by the British government to examine closely into the affairs of Massachusetts, reported respecting New England: "There be five ironworks which cast no guns."[43] A few years later he reported to the home government the existence of six. During the rest of the century the industry made little progress. Although many men had shown enthusiasm and had done all they possibly could to promote the production of iron, many factors contributed to prevent the development of the industry. Lawsuits, Indian attacks, prejudice and opposition because of the great quantity of wood used for the charcoal fuel, and a preference for English iron, were reasons that prevented the growth of the New England industry before 1700. Soon after the opening of the eighteenth century New England began to manufacture iron articles from iron received from Old England. In 1689, a duty of fifty-four shillings a ton had been imposed by Parliament on all foreign unwrought iron imported into England. In 1703, a drawback of fifty shillings and six pence was allowed on its re-exportation.[44] The New Englanders took advantage of the opportunity to import iron and to work it into manufactured goods. By 1710, English manufacturers began to complain and demanded that Parliament discontinue the drawback in order to stifle such manufactures in the New World. It was about this time that the iron industry took root in New England. By 1731, there were in that region, six furnaces that cast hollow-ware such as kettles, pots and pans, nineteen bloomery forges, one slitting mill and one nail works.[45] From this time on the industry developed and grew.

The first important promoter of iron manufacture in New Jersey was Colonel Lewis Morris, who in 1676 moved to New Jersey from Barbadoes. Almost immediately he and several associates received a grant of lands and the right "to dig, delve, and carry away all such mines for iron . . . to the iron work."[46] These works were located at Shrewsbury, near Tinton Falls, Monmouth county, and had been built about the middle of the century.[47] The bog ores of the region were used. Under Morris and his partners the works prospered, but little further was done in the manufacture of iron in this colony until the second decade of the eighteenth century, when the bog ores of southern New Jersey, and a little later, the rich ores of northern New Jersey, were utilized and ironworks were established in rapid succession.

Maryland set up the first bloomery in 1715,[48] about a year before the first forge was built in Pennsylvania. In 1718, samples of iron made at this forge, which became known as the Principio Forge and located at the head of the bay in Cecil county, were sent to England.[49] Beginning in 1719, acts were passed by the legislature of Maryland, which extinguished the rights of the Lord Proprietary to the royalty on minerals, and also authorized the gift of one hundred acres of land to any who would establish furnaces and forges there.[50] Following the building of the furnace of the Principio Company in 1724,[51] much progress was made in the manufacture of iron in Maryland due largely to legislative encouragement.[52]

Not a single ironworks was built in Pennsylvania until long after the English Quakers settled there. The first white settlers along the Delaware River were the Dutch and the Swedes. The people of these scattered settlements, however, made no attempts to manufacture iron. As early as 1623 the Dutch from New Netherland began to make settlements on the South (Delaware) River. They were not successful, but they established Fort Nassau, near the present site of Gloucester, New Jersey. This fort was established primarily as a trading post

Hay Creek Forge

and remained Dutch headquarters in this region until 1651. In 1638, the Swedes appeared on the Delaware River. They made a few scattered settlements along the river, laid out farms, began trading with the Indians for furs, and established Fort Christina (Wilmington). In 1642, Colonel Johan Printz, a soldier schooled in the Thirty Years' War, was appointed by the Royal Chancellor of Queen Christina as the third governor of New Sweden. Printz did not like the location of Fort Christina and therefore decided on Tinicum Island as the new site for his fort, settlement and mansion, because the island gave him excellent command of the river.[53] Although Old Sweden was rapidly developing an extensive iron industry at this time, and many Swedes were fired with the desire to emulate the mother country in regard to iron manufacture, in the New World they faced too many difficulties in establishing settlements to attempt the production of iron.

The Dutch of New Netherland were at first too busy fighting the Indians in several regions to pay much attention to the Swedish settlements on the territory claimed by them. Printz's colony flourished, but there was marked ill feeling between the Swedes and the Dutch. An agreement was made in 1648 between Governor Stuyvesant, the able governor of New Netherland, and leisurely Governor Printz, whereby each promised not to commit any hostile act against the other. However, in an attempt to control the river, Stuyvesant in 1651 built Fort Casimir, near what is now the town of New Castle, Delaware. Three years later, a Swedish vessel sailed up the river and took the fort. The next year, seven ships in command of Governor Stuyvesant, left New Amsterdam, entered the Delaware Bay and re-captured Fort Casimir.[54] This marked the end of the rivalry between the Swedes and the Dutch in the Delaware Valley, and the dreams of a New Sweden in America faded forever.

Dutch sway was short-lived. In 1664, without a word of warning, an English fleet attacked New Amsterdam. In spite of the protests and ravings of Stuyvesant, the entire region between the Hudson and the Delaware Rivers was seized by the English without opposition. Charles II granted this land to his brother, James, Duke of York.[55]

During these early attempts at settlement, conditions were not favorable to attempt the working of iron. The Dutch were quite interested in the discovery of minerals and metals at a very early period. In 1646, Andrie Hudde received instructions from William Kieft, director general of the New Netherlands, to discover the mineral properties in the regions of the Delaware River. The Swedish governor, Printz, however, incited the Indians against Hudde, and he was not allowed to proceed beyond the falls of the river, at Trenton, New Jersey.[56] The rivalries between the Dutch and the Swedes on the Delaware undoubtedly played a part in preventing the discovery of deposits of ore and possibly the manufacture of iron.

The first iron used in the Delaware settlements came from Europe and was worked by the Dutch and Swedish blacksmiths. Iron was needed in the form of horseshoes, tools, nails, hinges, pots and pans, and for many other uses. On Tinicum Island, during Governor Printz's administration, there was a large blacksmith shop which provided iron implements needed by the colonists. In his report to the West India Company in old Sweden, Governor Printz designated the various kinds of artisans that were urgently needed in New Sweden, including blacksmiths and gunsmiths.[57] No record has been left, however, of any attempt to utilize the loose ores of nearby regions, or the black magnetic iron sand, which existed in large quantities along the banks of the Delaware.[58]

Just before sailing from England for his new colony, in order to have direct access from his province, Pennsylvania, to the ocean, William Penn secured from the Duke of York the land which became known as the three lower counties or territories along the Delaware River.[59] In 1704, these counties set up a separate assembly, although they remained under the same governor as Pennsylvania until the Revolution.[60] A consideration of early iron manufacture in Pennsylvania must therefore include the attempts made to establish ironworks in the region which now constitutes the state of Delaware.

Even before he reached America, William Penn had written a pamphlet, in 1681, in which he mentioned iron among the commodities "that the country is thought capable of."[61] Again in 1685, Penn wrote that there was much iron and wood in Pennsylvania.[62] Although he was connected with iron manufacture at Hawkhurst, England,[63] and was anxious to see ironworks built in his colony, as is quite evident in his correspondence, Penn took no active steps to establish any himself. He tried to encourage others to erect such works, but without success. Just before the seventeenth century closed, Gabriel Thomas, a resident of Pennsylvania, wrote that the quality of iron found in the province was richer than that of England, and that preparations were being made to build ironworks.[64] The plan, however, did not mature. After Penn returned to England in 1701, he tried to secure the aid of Englishmen, especially Sir Ambrose Crowley, the famous Newcastle ironmaster, in the project.[65] Because interest was lacking in the colony, and as it was impossible to get English aid, nothing was accomplished at this time.[66] Penn himself, like many others, was more anxious to discover silver, lead and tin.[67] His correspondence shows at times that he was hoping to retrieve his wasted fortune in this way.

Another attempt, dating from the earliest days of settlement, to produce and manufacture iron in Pennsylvania also failed. As early as 1682, William Penn granted an exceedingly liberal charter to the Free Society of Traders.[68] The plans of the company in many respects ran parallel to those of the earlier Virginia Company of London. Men were sent in the first year of Quaker settlement from England to Durham, about fifty miles from Philadelphia. Many manufactures were planned, including the production of iron.[69] But the rich mines of Durham remained untouched until James Logan, the loyal friend of the proprietors, and others built works there about 1727, or a little earlier.[70]

Although all the early attempts to establish ironworks in Pennsylvania failed, increasing amounts of iron were used in shipbuilding from the earliest days of Quaker settlement. Chain-plates, rudder iron, as well as spikes and nails, were necessary in the construction of vessels. Iron imported from England in the form of bars, was used by the Philadelphia blacksmiths and other ironworkers for this purpose.[71] An early Pennsylvania poet-historian as early as 1689 felt moved to sing praises to this developing industry:

BUILDING SHIPS

Within this six or seven year
Many good ships have been built here,
And more, some say, will be built yet,
For here is timber very fit.
Good carpenters who bravely thrive,
And master builders to contrive
Who can prepare the iron stuff.[72]

The neglect of the great iron deposits of the colony in a search for precious metals was also pointed out. As in other colonies, with visions of securing great wealth, many sought gold and silver to the neglect of other metals:

MINES

Besides what is upon the land
Here's divers mines now come to hand
In many places some do find
Good iron, which they do not mind;
They hunt and search for richer stuff
By which they hope to make enough:
Some men without digging deep in
Have found and run both lead and tin:
And here is copper of the best
That ever came to the test
In shining it resembles Sol
For which some do it much extol,
So maulable that with much ease
The artificer, when he please,
Makes of it what he hath a mind
It works so pliable and kind.
But little progress yet is made
In any of this mineral trade;
Some small beginnings do go on
And more may be ere it be long.
Had we but men of wealth and skill,
Who would go on with a good will
In prosecuting of this same,
They might get wealth, honor and fame.
More might be said, but I have reason
To leave this till another season.[73]

While copper mines were discovered and worked at different times throughout the eighteenth century, copper was never found in Pennsylvania in very great quantities.[74]

It was natural that blacksmiths should be the first to attempt to work the iron ores, found loose on the surface of the earth, into much-needed iron implements and ware. Richard Frame, another early historian, writing in 1692, became lyrical in describing such an instance:

If men would venture for to dig below,
They might get well by it for aught I know:
Those Treasures in the Earth which hidden be,
They will be good, whoever lives to see.

25

A certain place here is, where some begun
To try some Mettle, and have made it run,
Wherein was iron absolutely found,
At once was known about Forty Pound.[75]

It proved quite profitable, although difficult, for blacksmiths to work iron from the rich ores during the early days of the colony. Gabriel Thomas, who was interested in securing immigrants for the new country, wrote that his neighbor in Philadelphia, with a Negro helper, earned fifty shillings in one day by working up one hundred pounds of iron.[76] Deputy Governor Keith, at the beginning of his administration, wrote to the Board of Trade that there was plenty of iron in Pennsylvania, and asked for aid in establishing ironworks. He complained that blacksmiths with their "common furnaces work up iron to great advantage and in such great quantities as thereby to discourage the importation and lower the price of European Iron."[77] Like Governor Spotswood in Virginia, Keith believed that the colonies should be encouraged to smelt iron for the mother country to manufacture into tools, utensils and other iron products.[78]

The method of working iron in a blacksmith's forge required the highest grade of ores, and as intense a blast as was possible to obtain from the small blacksmith's bellows whether worked by hand power or water power. The process was hard and tedious, and when successful, resulted in but a small amount of iron. Although iron ores were abundant, dense forests at hand to supply the fuel, and many streams and creeks to furnish water power, as well as a pressing need for iron tools, implements and utensils, more than thirty years passed from the time of Penn's first Quaker settlement before the first bloomery forge was erected.

FOOTNOTES

1 J. Doddridge, *Notes on the Settlement and Indian Wars of the Western Parts of Virginia and Pennsylvania*, p. 77.
2 W. K. Moorehead, *Hematite Implements of the United States*, p: 57; W. H. Holmes, "Traces of Aboriginal Operations in an Iron Mine Near Leslie, Missouri," Smithsonian Institution, Annual Report, 1903, pp. 723-726.
3 David Zeisberger, *History of the Northern American Indians*, pp. 28 ff; A. E. Douglass, "A Table of the Geographical Distribution of American Indian Relics," American Museum of Natural History, Bulletin VIII, 199-220.
4 *Journal of Columbus' First Voyage*, pp. 24 ff.
5 *Letters of Columbus and Other Original Documents*, *Publications of the Hakluyt Society*, Series 1, XLIII, 6, 31, 38.
6 J. B. Thacher, *Christopher Columbus*, *Life, Work, Remains*, pp. 569 ff.
7 W. H. Prescott, *History of the Conquest of Mexico*, I, 141-143.
8 Garcilasco Lasso de la Vega, *Commentarios Reales*, I, 299-301.

[9] Gonzalo Fernandez de Oviedo y Valdes, *Historia General y Natural de Las Indias*, IV, 213 ff.

[10] T. A. Herrera, *Historia General*, V, 90-91.

[11] W. K. Moorehead, *Prehistoric Implements*, pp. 17 ff.

[12] David Zeisberger, *History of the Northern American Indians*, pp. 28 ff.

[13] W. R. Birdsall, "The Cliff Dwellings of the Canons of the Mesa Verde," in American Geographical Society *Journal*, XXIII, 584-620.

[14] T. A. Herrera, *Historia General*, IV, 154-156; Peter Martyr, *The Decades of the Newe Worlde*, in Richard Eden, *The First Three English Books on America*, pp. 355 ff.

[15] F. de Fonseca and D. C. Urrutia, *Historia General de Real Hacienda*, I, 5-44; II, 297-387; III, 6-140; IV, 521-636; V, 43-57. A chapter on mines and mining may be found in H. H. Bancroft, *History of the Pacific States: Mexico.* VI, chap. XXVIII. See also H. I. Priestley, *The Coming of the White Man*, pp. 35, 39, 85-88, 93-95.

[16] Thomas Hariot, *A Briefe and True Report of the New Found Land of Virginia.*

[17] *Ibid.*, p. 17.

[18] George Barrington to the Board of Trade, Feb. 20, 1731-32, *North Carolina Colonial Records*, III, 337; A. Dobbs to the Board of Trade, Jan. 4, 1755, *Ibid.*, V, 316-317; *Ibid.*, VII, 898 ff. *passim;* Governor Tryon to the Earl of Hillsborough, Feb. 10, 1769, *Ibid.*, VIII, p. 166; Governor Tryon to the Earl of Hillsborough, Feb. 8, 1771, *Ibid.*, VIII, 496.

[19] Nova Brittania (1609), in Peter Force, *Tracts and Other Papers*, I, No. 6, p. 16; *A True and Sincere Declaration of the Purpose and Ends of the Plantation* (British Museum), in Alexander Brown, *Genesis of the United States*, I, 340, 353, 356.

[20] *Calendar of State Papers, East Indies*, 1513-1616, p. 181.

[21] Letters of the Virginia Company, Edward D. Neill, *History of the Virginia Company*, pp. 239, 270, 283.

[22] *Ibid.*, pp. 302 ff.

[23] *Ibid.*, p. 338.

[24] *Records of the Virginia Company of London*, II, 469, 473-474. For an account of Spotswood's early enterprises, see A. C. Bining, *British Regulation of the Colonial Iron Industry*, pp. 20-21.

[25] Board of Trade Journals, XXXI, 342-343. Official Letters of Governor Spotswood, 1710-21, Virginia Historical Society, *Collections*, New Series I, II.

[26] *Records of the Governor and Company of Massachusetts Bay*, I, 28, 30, 32-33.

[27] John Winthrop's Draft of Petition to Parliament, Winthrop Papers, Massachusetts Historical Society, *Collections*, Fifth Series VIII, 36-37; see also Massachusetts Historical Society, *Proceedings*, Second Series, VIII, 14; *Ibid.* Fifth Series, VIII, 36, 37. John Winthrop, *History of New England*, II, 212.

[28] *Records of the Governor and Company of Massachusetts Bay*, II, 61.

[29] *Ibid.*, p. 103.

[30] Lynn Iron Works MSS.; "Winthrop Papers," Massachusetts Historical Society, *Proceedings*, Second Series VIII, 13. ff. The Lynn Iron Works were often called the Hammersmith Iron Works because the skilled workers were brought over from Hammersmith, England. For the history of these works see A. Lewis and J. R. Newhall, *History of Lynn*, pp. 93-280, *passim.* Essex Institute, *Collections*, XVIII, 241 ff.; W. S. Pattee, *History of Old Braintree and Quincy*, pp. 450 ff.

[31] *Records and Files of the Quarterly Courts of Essex County, Massachusetts*, I, 284 ff. *passim;* II, 130 ff. *passim;* III, 42 ff. *passim;* IV, 150 ff. *passim;* V. 44 ff. *passim;* VI, 166 ff. *passim;* VII, 26 ff. *passim;* VIII, 38 ff. *passim.*

[32] Massachusetts Historical Society, *Collections*, Second Series, IV, 224.

[33] *Ibid.*, First Series, III, 170 ff; F. Baylies, *Historical Memoir of the Colony of New Plymouth*, I, Part II, 268; W. R. Deane, *Genealogical Record of the Leonard Family*, pp. 5 ff.

[34] *Records of the Governor and Company of Massachusetts Bay*, IV, Part I, 311.

[35] *Records and Files of the Quarterly Courts of Essex County, Massachusetts*, VI, 1-5 *passim;* J. B. Felt, *Annals of Salem*, I, 282.

[36] *Records of the Governor and Company of Massachusetts Bay*, I, 223; Petition in Winthrop Papers, Massachusetts Historical Society, *Collections*, Fourth Series, VI, 517-518.

[37] Benjamin Trumbull, *Complete History of Connecticut*, I, 169 ff.

[38] William Paine to John Winthrop, Jr. 26th 11th month, 1657, in Winthrop Papers, Massachusetts Historical Society, *Collections*, Fourth Series, VII, 403.

[39] J. S. Palfrey, *History of New England*, II, 234 ff.

[40] Roger Williams to John Winthrop, 15 12th month, 1654, in Winthrop Papers, Massachusetts Historical Society, *Collections*, Fourth Series, VI, 286-292.

[41] *Records of the Governor and Company of Massachusetts Bay*, III, 65, 386; IV, Part I, 233; IV, Part II, 348, 528.

[42] Alonzo Lewis, *History of Lynn*, p. 121.

[43] A. Holmes, *American Annals*, I, 416.

[44] 1 and 2 William and Mary, c. 4; 2 and 3 Anne, c. 9.

[45] William Douglass, *British Settlements in America*, I, 540; II, 109.

FOOTNOTES—Concluded

46 *Papers of Lewis Morris*, pp. 2, 3; George Scot, *Model of the Government of the Province of East New Jersey*, p. 271; Samuel Smith, *History of the Colony of Nova-Caesaria or New Jersey*, p. 430; W. A. Whitehead, *East Jersey Under the Proprietary Governments*, I, 91.

47 E. B. O'Callaghan, *History of New Netherland*, II, 342.

48 Adam Anderson, *Origin of Commerce*, III, 63; Harry Scrivenor, *History of the Iron Trade*, p. 68.

49 *Ibid.*, Appendix, p. 328.

50 *Maryland, Acts of Assembly*, 1719, c. XV; 1732, c. VII; 1736, c. XVII.

51 An excellent sketch of the Principio Company is that by W. G. Whitely, "The Principio Company," *Pennsylvania Magazine of History and Biography*, II, 63-68; 190-198; 288-295. (The article credited to W. G. Whitely was written by Henry Whitely.)

52 J. H. Alexander, *Report on the Manufacture of Iron Addressed to the Governor of Maryland*, pp. 70 ff. For a survey of the entire colonial iron industry see A. C. Bining, *British Regulation of the Colonial Iron Industry*, Chap. I.

53 Israel Acrelius, *History of New Sweden*, pp. 73 ff. In Christoph Daniel Ebeling's *Delaware*, in *Die Vereinten Staaten von Nordamerika*, the fifth volume of *Erdbeschreibung und Geschichte von Amerika*, is a good account of the history of New Sweden based on almost all the works then published. (1799.)

54 Israel Acrelius, *History of New Sweden*, pp. 75 ff; Amandus Johnson, *Swedish Settlements on the Delaware*, I, 117, 170.

55 *Documents Relative to the Colonial History of the State of New York*, II, 272, 273, 275, 276, 281.

56 Report of Andrie Hudde, in New York Historical Society, *Collections*, Second Series, I, 428; *Documents Relating to the History of the Dutch and Swedish Settlements on the Delaware River*, XII, 28.

57 Report of Governor Printz to the Right Honorable West India Company. A copy of this report may be found at the Historical Society of Pennsylvania. It has been translated and published in *Pennsylvania Magazine of History and Biography*, VII, 271 ff.

58 Israel Acrelius, *History of New Sweden*, p. 169.

59 Copy of Deed, Duke of York to William Penn, August 24, 1682, in Penn MSS: Charters and Frame of Government, Documents 5-7.

60 *Pennsylvania Colonial Records*, II, 116, 119, 120.

61 William Penn, *Some Account of the Province of Pennsylvania*, p. 5.

62 William Penn, *A Further Account of the Province of Pennsylvania*, p. 14.

63 Samuel Smiles, *Industrial Biography*, p. 56.

64 Gabriel Thomas, *Account of the Province of Pennsylvania*, p. 28.

65 William Penn to James Logan, 21st, 4th month, 1702, *Penn Logan Correspondence*, I, 115.

66 James Logan to William Penn, 1st 10th month, 1702, *Ibid.* I, 149.

67 William Penn to James Logan, 3, 3rd month, 1708, *Ibid.* II, 269 ff. See also pp. 295, 297, 314-315, 323.

68 A copy of the charter may be found in Samuel Hazard, *Annals of Pennsylvania*, p. 542; see also Samuel Hazard, *United States Commercial and Statistical Register*, I, 394; *Pennsylvania Colonial Records*, II, 136, 153, 160, 163; III, 138.

69 *Penn-Logan Correspondence*, June 4, 1682, II, 323.

70 James Logan to John Penn, December 6, 1727, in Logan Papers, MSS., I, Document 89; Penn MSS. Indian Affairs, 1757-1722; p. 105. There is evidence to show that iron was manufactured at Durham, possibly, in a bloomery before 1727 when the blast furnace was built.

71 Board of Trade Papers: Plantations General, IX, K 7.

72 John Holme, *A True Relation of the Flourishing State of Pennsylvania*. The original, presented to the Historical Society of Pennsylvania, has been lost. Before its loss it was printed in the *Bulletin* of the Historical Society of Pennsylvania, I, 161 ff.

73 *Ibid.*, p. 171.

74 C. D. Ebeling, *Die Vereinten Staaten von Nordamerika*, IV, 89 ff.; S. W. Pennypacker, *Henry Pannebecker*, pp. 76 ff.

75 Richard Frame, *Short Description of Pennsylvania*, p. 4.

76 Gabriel Thomas, *Account of the Province of Pennsylvania*, p. 28.

77 Board of Trade Papers: Proprieties, X (1), Q 140.

78 Board of Trade Papers: Journals, XXXI, 342-343.

CHAPTER II

THE IRON PLANTATIONS

SCATTERED here and there over southeastern Pennsylvania, especially in the Schuylkill Valley, trailing through the wide Susquehanna Valley, along the beautiful blue Juniata, and across the wooded Alleghenies, may still be found the ruins of old furnaces. Each ruin—a pile of large stones intertwined with leaves and the wild growth of bramble—was once the scene of great activity, the center of a community where the ironmaster and his dependents lived and labored. Here the pioneer ironworkers of Pennsylvania, toiling hard, produced iron needed for manifold purposes, and played their part in laying the foundation of a great Commonwealth. Although most of these communities, or "iron plantations," had their origins in the eighteenth century, many remained until the middle of the nineteenth, and even later. With the development of large capitalistic enterprises and industrial consolidation after the Civil War, they gradually disappeared and became mere memories.

In many places where iron plantations once flourished, nothing now remains except the stately old mansion house. Those who live there today can usually point out the site or the ruins of the old furnace or forge, and the nearby "cinder" or "slag" heap, buried beneath a layer of soft earth and leaves. Such mansion houses as Reading (Redding), Warwick, Coventry, Stowe, Windsor, Hopewell, Pine, Pool, Spring Grove, Boiling Springs, Elizabeth, and many others, stand as monuments to a race of fearless ironmasters who faced tremendous difficulties in obtaining capital, securing skilled workmen, and dealing with metallurgical problems in an age of experimentation.

Scarcely a trace is left of the many forges which once prospered. As most of them were built of wood, time has taken her toll and few remain. One or two, built of stone, still stand, but the old water-wheels, forge hammers, and hearths, are gone. Within those walls the forging of iron has given way to the grinding of grain, or to the sawing of wood. Most of the houses or log cabins, where the ironworkers lived and died, have also disappeared.

During the eighteenth century, the iron industry in Penn-

Mansion House, Warwick Furnace

sylvania was organized largely on plantations. Many of these consisted of several thousand acres of land. The mansion house, the homes of the workers, the furnace and forge or forges, the iron mines, the charcoal house, the dense woods which furnished the material for making charcoal, the office, the store, the gristmill, the sawmill, the blacksmith shop, the large outside bake oven, the barns, the grain fields, and orchards, were part of a very interesting and almost self-sufficing community. In some respects the iron plantations resembled small feudal manors of medieval Europe.

It is not strange that these plantations were of great extent. Much woodland was necessary for the thousands of cords of wood consumed annually in the form of charcoal by each furnace. Forges used much less, but still a relatively large amount of fuel. Many acres were also needed for cultivating grain and raising the food required by the workers. Thus, after Samuel Nutt began the manufacture of iron on French Creek, as the forests were cut down, he added thousands of acres to his vast estate.[1] Elizabeth plantation consisted of 10,124 acres,[2] Durham of 8,511 acres, Boiling Springs of 7,000 acres,[3] Reading of 5,600 acres,[4] Martic of 3,400 acres,[5] Warwick of 1,796 acres,[6] Pine of 1,280 acres,[7] and Thornburg of 1,200 acres.[8] During the latter part of the eighteenth century, Colebrook plantation comprised 7,684 acres,[9] and the tract upon which Cornwall Furnace and Hopewell Forges stood, extended over 9,669 acres.[10] A few were much smaller, such as Mount Joy (Valley Forge) plantation, which about the middle of the century included only 375 acres,[11] and Coventry, which in 1771 consisted of 600 acres.[12] The iron plantations organized in western Pennsylvania during the last two decades of the century, were in general much smaller than those of the older southeast. The Alliance Iron Works estate consisted of only a few hundred acres.[13] On the smaller plantations, the ironmasters were forced to buy most of their wood from farmers and from the owners of woodland in the neighboring districts.[14]

The mansion house, where the ironmaster and his family lived, was usually built on a low hill overlooking the furnace or forge. In the surrounding flower garden, lilies, violets, pinks, hollyhocks, phlox and many other old-fashioned flowers bloomed. The house was stately and commodious, with large rooms, having wide, open fireplaces, and furnished with excellent furniture often imported from Europe. In the kitchen, the large kettle, suspended from a crane over a wood fire, the candlesticks and glasses on the mantel-shelf above; and the pewter plates, china dishes and brown earthenware in the corner cupboard, gave a very homelike appearance. The other downstairs rooms contained oak or mahogany furniture including damask-covered couches. China cups and saucers, imported Delft ware, and tankards were carefully arranged on shelves. Fires were kept burning brightly in the huge fireplaces during the long winter evenings. The upstairs rooms, especially the guest room, contained curtained beds, massive chests of drawers, and other heavy furniture of the period.

The houses of the ironmasters, built in the west during the

beginnings of the iron industry there, were not quite as pretentious as those of the older east, nor were they so well furnished. Yet, built only of wood and plaster, these homes stood in marked contrast with the crude workmen's cabins which surrounded them.[15] The large stone house erected by Nathaniel Gibson on his iron plantation at Little Falls, Fayette county, was an exception and was one of the first large and commodious mansion houses constructed in the Monongahela country before 1800.

Mansion House and Store, Sally Ann Furnace

In the east, the cottages of the furnacemen, forgemen, miners, farm-hands, and other workers, were usually small stone structures, or were built of log and plaster with stone chimneys.[16] In the central and western parts of Pennsylvania, they were more often log cabins.[17] All were poorly furnished. Until the end of the eighteenth century, rugs and carpets were unknown in the homes of the workers. Sanded floors and whitewashed walls often made for cleanliness. In the smaller houses, there were but two rooms with a loft above. In the better type of workingmen's homes were an additional room or two. Cooking was done at the fireplace which also provided heat during the winter months. In only a few homes could a cast-iron stove be found. Pewter dishes, plates and spoons; iron knives and forks; and wooden bowls and trenchers were the utensils used in these homes at mealtime. The bedrooms were bare and rarely contained mirrors, tables, wardrobes, drawers, or even chairs.[18]

Just below the mansion house and not very far away from the dwellings of the workmen stood the furnace, a truncated pyramid of stone. Built into the side of a small hill in order that the ore, limestone flux, and charcoal could be put into the furnace at the top, it was an impressive sight when in blast. The intermittent roar of the forced blast could be heard a long distance away. From the top of the furnace stack a

Old Buildings, Warwick Furnace

stream of sparks was occasionally emitted as the flames rose and fell. At night the almost smokeless flames cast a lurid glare upon the sky, visible for miles around, which illuminated the surrounding buildings. Within the main casting house or casting shed as it was called, which was built directly in front of the furnace, the "mysteries" of casting were carried on. Here the molten metal was run from the hearth into the waiting molds of scorched and blackened sand. Creaking wagons drawn by teams of horses hauled the iron ore up the furnace road. From the "bank," the fillers carried their baskets of ore, limestone and charcoal across the bridge to the furnace top. Pig iron was the chief product of the blast furnace, although pots, pans, kettles, stove-plates and fire-backs were also cast.

The forge, where the pig iron was refined and hammered into blooms, or bars of wrought iron, was generally not far distant. The dull, unvaried turning of the water-wheel, the irregular splash of falling water, the rhythmic thump of the hammer, and the droning sound of the anvil, were a part of life on the plantation. Within the forge half-naked human beings of strong physique, swung the white-hot pasty metal from the hearths to the great hammers by means of wide-jawed tongs. Under the steady strokes of the hammers, amid showers of scintillating sparks, the forgemen drew the bar to given sizes. Bar iron from the forges was used by blacksmiths to make tools, implements, and ironware of different sorts.

In the midst of the community was the ironmaster's store. All the necessities of life needed by the workmen and their families were secured there. It resembled in many ways a large country store of the present time, carrying supplies of all kinds. Grains and vegetables raised on the plantation, flour ground at the mill, the meat of cattle and animals bred in the fields, axes, shovels, chains, and hoes made by the plantation blacksmith, sugar and molasses from the West Indies, rum from New England, and imported goods from old England could be purchased. Broadcloth, linen, flannel, and other varieties of cloth were kept, as well as shoes, deerskins, scythes, fireplaces, stoves and castings.[19] The workers depended upon the store for drugs and medicines. Jesuits' bark or Peruvian bark and medicines called "purges" and "vomits" were obtainable. Even coffins and funeral supplies were bought at the store.[20] The store was a necessity because of the distance to the boroughs and also because of the scarcity of money.

On the stream far below the furnace or forge was the gristmill, built of logs, thick boards or stone. From the high gable, a rope like a hangman's, for raising grain swung in the breeze. On one side of the building the ponderous wheel, dark and green with slimy moss, was driven around slowly by the stream of water. The ceiling of the mill was covered with cobwebs and the walls whitened with meal dust. The sound of grinding that issued forth was soft and low, for the machinery was all made of wood.

Almost all the iron plantations possessed a sawmill. Timber had to be prepared for erecting buildings and for other purposes.[21] At the beginning of the eighteenth century, there seemed to be enough wood in the dense forests to last for ages. However, before the middle of the century, the traveler, Peter

Kalm, commented on the high prices of lumber in the neighborhood of Philadelphia and predicted a shortage in the future.[22] This was due not alone to the large quantities of staves, shingles, planks and boards which were exported,[23] but also to the vast amount of wood consumed by furnaces and forges.[24] This problem did not affect the west at this time, because in the Alleghenies and beyond the frontier lay vast forests.[25]

Ironmaking was only a part of the work on the plantations. All the cereals—wheat, buckwheat, corn, rye, oats and barley were grown.[26] Plowing was mostly done with horses; occasionally oxen were used.[27] The system of cultivation was poor. Land was sown with wheat until it would bear wheat no longer, then with barley until barley would no longer thrive, followed by oats, buckwheat and peas. The exhausted land was forsaken and a new piece of ground cleared.[28] Flax and hemp were cultivated to some extent and sheep were raised, chiefly for wool.[29] In southeastern Pennsylvania, however, several decades before the close of the eighteenth century, efforts were made to improve agriculture and stock raising. Men like Lynford Lardner were interested in the movement for agrarian reform that was taking place in England and applied such knowledge as they learned to their own farms. At the Warwick iron plantation and at others long before the century closed, the rotation of crops was practiced, lime was used as a corrector of soil acidity, and experiments were made in fertilization. The Philadelphia Society for Promoting Agriculture offered many premiums for agricultural improvements.[30] Other societies were organized, such as the Blockley and Merion Society for Promoting Agriculture and Rural Economy.[31]

While the workmen toiled at furnace, forge, or mine, the women in the homes spun thread and wove cloth on spinning wheels and looms. In haytime and harvest, the women and children turned out to work long hours in the fields. Days were appointed for husking bees, in which old and young participated. The women even did the harder farm work also. They made hay, pulled flax and turnips, helped at threshing, reaped the grain, and in fact did almost all the harvesting.[32]

The workers handled little money. The English, French, German, Dutch, Portuguese and Arabian gold and silver coins,

35

the bills of credit and other paper currency used in Philadelphia and occasionally in the scattered boroughs and villages,[33] were rarely seen on the plantations. The ironmaster credited his workers with the amount earned by them each day on one side of the ledger; on the debit side he made entries of merchandise and goods bought at the store. Likewise the ironmaster often received in exchange for iron sent to Philadelphia, barrels of rum, dozens of pairs of shoes, and other merchandise, which he placed in his store to be sold to his workmen.[34]

As most of the iron plantations were some distance from the boroughs, settlements and villages, the workers did not travel far. On many of the isolated plantations, especially during the first half of the century, they knew little of what was happening in the world outside. All their interests were bound up in their community. Not often did many of them have the opportunity even to see the stage coaches and stage wagons which ran from Philadelphia to such centers as New York, Baltimore,[35] Harrisburg, Reading, and Bethlehem.[36] As time went on and some of the regions of Pennsylvania were built up, this isolation tended to disappear. The iron plantations in the Schuylkill Valley, for instance, by the end of the eighteenth century, were in a fairly populous district.[37]

The ironmaster often employed a tutor or an old schoolmaster to teach his children.[38] The children of the workers did not have the opportunity of receiving the smallest amount of training, although occasionally, those of the better paid workmen obtained instruction from the schoolmaster.[39] Frequently, the clerk or bookkeeper at the plantation store, being fairly well educated, served as the teacher. The inadequate number of church schools, or even the neighborhood schools which developed later,[40] were too far away from the plantations to be of service to those who lived there. This was true not only in the west, but also in the more settled southeast for most of the period.

Churches were too few in all parts of Pennsylvania during the eighteenth century.[41] If five or ten miles away, church could be attended occasionally by traveling there on horseback or in wagons. Some ironmasters, like "Baron" Henry William Stiegel at Elizabeth, conducted their own sacred services.[42] Many itinerant preachers called at the different plantations. According to tradition, the famous George Whitefield preached at Warwick. The rough ironworkers threatened to kill him, but his life was saved by the appearance of Mrs. Robert Grace, the ironmaster's beautiful wife, who became a Methodist herself. She allowed one of the buildings on the plantation to be used as a chapel.[43] Years later, Benjamin Abbott, a follower and imitator of the great preacher, wrote of his visit to Warwick in 1780. He stated that Mrs. Grace, who was now old, sent a person to the meeting when he preached at Warwick to prevent the ironworkers from killing

Old Office Building, Warwick Furnace

him, and that he was shown every courtesy by her.[44] John Cuthbertson, the first Reformed Presbyterian minister to arrive in America (1751) traveled on horseback through the forests from one community to another. He visited Reading (Redding) Furnace, Warwick, Codorus, and others. Not only did he preach and comfort his people, but he also butchered their cattle at the various places where he called.[45]

During the period under discussion, most of the inhabitants of Pennsylvania drank much liquor and strong drink.[46] The ironmasters could afford Sherry, Catalonia, Madeira, French and Teneriffe wines, Jamaica spirits, Bordeaux claret, and bottled porter.[47] The workers on the plantations drank much rum, whiskey, gin, cider, and beer.[48] Drunkenness among the furnace workers was common. It was for this reason that the legislature passed acts in 1726,[49] and 1736,[50] prohibiting the sale of liquors near the furnaces. By the first act, no public house could be licensed within two miles of a furnace unless the ironmaster gave special permission at the time application was made for the license. The act of 1736 increased the radius to three miles. There was always a shortage of founders and skilled workers, and it was imperative that the furnaces be attended continually, or serious damage might result from the furnace running cold, from explosions, or from other mishaps. The first act stated that the "selling of rum and other strong liquors near the furnaces" had "proved very prejudicial and injurious to the undertakers."[51] This legislation gave the ironmasters the right to sell liquors to their own employees. In this way, they could regulate the sale of such beverages and at the same time secure a monopoly of its distribution within the area of the plantation and even outside its limits. The large consumption of liquors continued throughout the eighteenth century and into the nineteenth.[52]

While the lot of the workers on the plantations was one of toil, they found some amusement and relaxation in occasional barn dances, corn huskings, and country parties. Once or twice a year some of them were fortunate enough to be able to travel to the fair held at the nearest borough.[53] By virtue of their charters, the boroughs were allowed to hold a market each week and two fairs a year. The fairs became centers of attraction. General merchandise, supplies and live stock were sold. Many people attended them and there was much excitement and pleasure. Some went to make purchases; others planned a wild frolic. Horse racing, drinking and gambling often prevailed. At Bristol in 1773, the council resolved that the fairs were useless on account of the large number of shops and stores in the borough, and because "debauchery, idleness and drunkenness, consequent on the meeting of the lowest people together is a real evil and calls for redress."[54] It urged the legislature to abolish them. This was done in 1796.[55] Other boroughs had the same experience. In York, Lancaster and Harrisburg the fairs were abolished in 1816.[56] Not long afterwards they disappeared from all the boroughs. Agricultural and mechanical fairs took the place of the borough fairs which were a remnant of medieval days.

The highways which led from the plantations to the outside world and connected the boroughs, towns and villages were picturesque. Quaint signboards of taverns along the road, and of inns and shops in the boroughs and villages added to the attractiveness of the natural scenery. The Three Crowns, King's Head, Grayhound, Black Horse, Unicorn, Golden Eagle and Swan were but a few.[57] Every craftsman and shopman had his sign representing his calling.[58] After the Revolution many changes were made in the tavern signs. The British Union Jack became the Flag of the Thirteen United States of America, while perhaps to save the cost of painting, the Golden Lion was changed to the Yellow Cat.[59]

Along these roads and highways long before the middle of the eighteenth century, the famous covered wagons, which later played so great a part in the westward movement, were transporting merchandise, goods and produce, to and from Philadelphia, and between the boroughs and towns.[60] These "freight wagons," or Conestoga wagons as they came to be called, were strongly built, the body sloping forward very slightly. They were covered with coarse cloth stretched over hoops. More than 7,000 were in use by 1750.[61] Pig iron, castings, and bar iron were hauled from the furnaces and forges in heavy open wagons over tortuous roads to the main highways. The cost of transportation under these conditions was exceedingly high. For instance, the cost of taking one ton of pig iron which sold for £5-0-0 from Colebrookdale Furnace to Philadelphia, a distance of about forty miles, varied from £1-0-0 to £2-0-0.[62]

Until long after the close of the eighteenth century, the roads west of Carlisle were extremely poor. Packhorses carried all merchandise along Indian trails, centuries old, to the frontier. By the end of the century, as industry marched westward, bar iron sent from the Juniata regions to the Monongahela country was bent into the shape of the letter "U" to fit the backs of horses. These horses were led over the mountains in divisions of twelve or fifteen. Each horse carried about two hundred-weight. They traveled in single file in charge of two men, one leading and the other at the end.[63] For many years, Mercersburg, Franklin county, was an important center of trade with the Indians and the western frontiersmen.[64] As the century advanced, ironworks were built farther and farther westward, following the trail of the moving frontier.

By the close of the eighteenth century iron plantations had been established as far as the western borders of the state, and even beyond.[65] These did not supply the west with all the iron which was needed and for many years after, pack-horses plodded wearily over the mountains bearing their loads of bar iron.

Those plantations which were not far from the rivers could send their iron and produce to market by water. The Schuyl-kill River was used to some extent, but difficulties in navigation prevented its extensive development as a waterway during this period.[66] Iron was sent from Durham down the Delaware River to Philadelphia in boats first designed by Robert Durham about 1750.[67] Washington used these boats when he crossed the river with his depleted army at the crucial time in 1776 amid the snows and ice of winter.[68] The Susquehanna River was also used to some extent. Pig iron, bar iron and castings from the ironworks as far west as the Juniata Valley were shipped down the Juniata River to Middletown on the Susquehanna.[69] From that point the iron was sent by road to Philadelphia, or down the Susquehanna to Baltimore where some of it was exported to Great Britain and to the West Indies.[70] Nor was iron the only commodity sent from the iron plantations to Philadelphia or Baltimore for export. Furs and skins were often shipped with the iron.[71] By the last decade of the century, many improvements of the roads and waterways of the state were planned.[72]

Of the many European travelers who journeyed through Pennsylvania during the eighteenth century, only a few visited the iron plantations. This was chiefly because these communities were usually a great distance from the main highways. They were established in the more remote regions because of nearness to ores and vast woods. In 1734, Emanuel Swedenborg, Swedish preacher, theologian and philosopher, wrote of the ironworks on the Christiana River (New Castle county) and of the first ironworks erected in the Schuylkill Valley by Thomas Rutter and Samuel Nutt. His account contains the earliest description of the Pennsylvania ironworks.[73] Before the middle of the century, Peter Kalm, Swedish naturalist and traveler, writing of his travels, noted Crum Creek Forge, which was not far from the main highway to Baltimore.[74] Israel Acrelius, sent to the Swedish parish of Christina on the Delaware River as minister in 1749, and who remained there until 1756 when he returned to his native land, had the opportunity of visiting many of the Pennsylvania iron plantations, especially those of the Schuylkill Valley.[75] Dr. John D. Schoepf, a surgeon who accompanied some German troops to this country during the American Revolution, saw and wrote about many of the ironworks in southeastern Pennsylvania.[76] In the western country, Michaux called at Probst's Works or Westmoreland Furnace. Writing about his adventures, he stated:

> They directed me at the foundry which road I was to take, notwithstanding I frequently missed my way on account of the roads being more or less cut, which lead to the different plantations scattered about the woods.[77]

One of the few native Pennsylvanians who liked to travel in her own land was Elizabeth Drinker. Leaving Philadelphia on September 3, 1764, with her husband, she visited Durham Iron Works. In her diary she wrote:

> Left home after dinner—second day; B. Booth on Horseback, and his man Robert; H. D. and E. D. in the Chaise. Drank tea at the Red Lion, 13 miles from Phila.; lodged at Alexr. Brown's, 28 miles from town, good accommodations. Breakfasted there ye 4th, then went to James Morgan's at Durham Ironworks—48 or 50 miles from home. Roads very bad; stayed there to dinner; walked to the Furnace, where we saw them at work casting iron bars, &c.[78]

In 1777, the gossipy Jacob Hiltzheimer, also paid Durham a visit. In his Diary he wrote:

> Reached George Taylor's at Galloway's Iron Works [Durham] where we had everything we could desire.[79]

After the Battle of the Brandywine in 1777, George Washington stopped at Reading Furnace in Chester county[80] while many of his men were sent to Warwick, not very far away.[81] At this time these furnaces and many others were busy casting cannon and cannon balls for the Continental army. Washington went into winter quarters at Valley Forge in 1777-1778. The famous forge there was burned a few months before the American army arrived to spend the winter.[82]

Many legends and traditions have grown up around the iron plantations. The Legend of the Hounds, put into verse by George H. Boker[83] is one of the most dramatic and interesting. The setting of the story is the Colebrook Furnace in Lebanon county, although many claims have been made that it originated at several other furnaces. According to the legend, one of the early ironmasters was a man of violent temper, a heavy drinker, and exceedingly cruel, who ruled over

Mansion House, Durham Iron Works

his community like a despot, feared by all. One day he returned from a fox hunt, enraged and cursing because the hounds had played him false. He drove the entire pack up the furnace road to the open blazing tunnel head. With whip in hand, he forced them one by one into the flames until only his favorite dog remained, quivering with fear at his side. Picking her up he made a motion as though he were about to cast her into the furnace. With terror in her eyes, she licked his hand. But he hurled her too into the furnace with a curse. A low, fearful moan escaped her and it was all over. The inhuman ironmaster never hunted again after that. Tortured with gout and with senses dulled with strong drink, he sat day after day before his open fireplace. He seemed to have no further interest in life. One morning he failed to appear. His servants found him seated upright in bed—dead, his hunting whip in his hand and his eyes set in terror. In the years that followed, many of the workers who lived in the neighborhood testified that on stormy autumn and winter nights they had heard the dread baying of hounds, and had seen the ironmaster flee in terrible fright before them.

Whether or not there is any foundation for this legend in the character of the ironmaster at Colebrook, most of the ironmasters were kindly men who took a real interest in the everyday life of their dependents. Thomas Rutter, the first to establish ironworks in the province; Samuel Nutt, another pioneer of industry; Robert Grace, a close friend of Benjamin Franklin; Thomas Potts and his sons and descendants who became prominent and influential in iron manufacture; the Birds of Birdsboro; the erratic Henry William Stiegel, the outstanding early German ironmaster; the Colemans; Mayburys; Olds; Grubbs; and many others, were all men who had much sympathy for their workmen and treated them well.

A legend, similar to the one told above, centers around Jacob's Creek Furnace in the western part of the state. Peter Marmie, a Frenchman and former private secretary to Lafayette, was one of the owners of this furnace. After a few years of prosperity, he met with business reverses. After his failure, according to the story, he called his hounds to him, assembled them on the bridge which led to the mouth of the furnace, and with whip in hand forced them into the blazing fire below. When the last one had disappeared, Marmie himself rushed headlong into the inferno. According to the story, the fires died out and were never rekindled. The stack went to ruins, the once prosperous community disappeared, and desolation marked the place where activity had abounded.[84]

It is true that many of the ironmasters kept packs of hounds and loved to hunt the fox. The hills around the plantations often echoed the hunter's horn and responded to the baying of his hounds. Many ironmasters imitated the life of English gentry. In the New World, there was more isolation, perhaps, than in England, but the Americans enjoyed the advantages of hunting, fishing and building fortunes that only a virgin country could give.

Life on the iron plantations, however, was cheerless at its best for those who had to toil hard and long. While the iron-master could secure many luxuries, could afford to travel,

Ruins of Alliance (Jacob's Creek) Furnace

even to Europe occasionally, and was able to provide teachers for his children, the lot of the workers was indeed very hard. There were few material comforts of life. It was a time when houses were lighted with tallow candles, when water had to be carried from springs or drawn from wells, when only a few had the opportunity of even an elementary education, when medicine as commonly practiced was a formulated superstition, rather than a science, and doctors purged and bled their patients. The few who traveled did so on horseback or in jolting stage coach or springless stage wagon, over rough limestone or poor clay roads. On the other hand, stark poverty was unknown on the iron plantations in spite of periods of occasional depression. The wages of the skilled workers were relatively high when compared with wages paid in European countries of the same time.[85] Initiative, industry and aggressiveness reaped rewards, for a number of the ironworkers by hard work and constant application became ironmasters themselves.

While the early furnaces and forges were organized on plantations, most of the other types of ironworks were not. Slitting mills at which was produced slit-iron for making nails; plating mills where bar iron was hammered into sheet iron or tin-plate iron; steel furnaces where small amounts of blister steel were produced for making tools; and air furnaces, the progenitors of modern cupolas, were usually built in towns or boroughs. A few of these, however, could be found on plantations, such as the slitting mill on the Brandywine,[86] and the steel furnace at Coventry.[87] But pig iron and bar iron were made on the iron plantations. These plantations originated during the colonial period. Beginning in southeastern Pennsylvania and spreading westward to the Monongahela country, they played an important part in the history and development of Pennsylvania long before Pittsburgh became a great iron and steel center.

The close connection between ironmaking and agriculture during the eighteenth and part of the nineteenth centuries contrasts strangely with the industrial organization of the present day. Many changes have taken place in the ironworks. The small stone furnaces have given way to the modern blast furnaces with their towering height of 100 feet or more, their four huge heating stoves, their blowing engines which deliver thousands of cubic feet of blast each minute, the array of dust arresters, gas washers, and automatic ore and coke handling machinery which are essentials of the giants of modern metallurgical devices. The forges at which iron was hammered out have been surpassed by rolling mills of various kinds. Water power has been superseded by steam and electric power, while coke as a fuel has taken the place of charcoal and also of anthracite, which was used to some extent after the third decade of the nineteenth century. The charcoal iron plantations have disappeared and have become memories. The pioneer ironmasters and ironworkers of Pennsylvania, however, played their part in establishing the foundations of a great industry which today makes the Pittsburgh district one of the greatest iron and steel centers in the world.

FOOTNOTES

1 *Pennsylvania Archives,* Third Series, I, 26, 55, 75.
2 *Pennsylvania Gazette,* December 5, 1765.
3 *Ibid.,* September 7, 1769.
4 *Pennsylvania Archives,* Third Series, XII, 65.
5 *Pennsylvania Gazette,* January 5, 1769.
6 *Pennsylvania Archives,* Third Series, XII, 295.
7 Will of John Potts, Mrs. T. P. James, *Memorial of Thomas Potts, Jr.,* p. 65.
8 Cumberland County Tax Lists, 1769.
9 *Alden's Appeal Record,* Coleman *vs.* Brooke, p. 88.
10 *Ibid.,* p. 30.
11 *Pennsylvania Gazette,* August 4, 1751, September 26, 1751.
12 *Pennsylvania Archives,* Third Series, XI, 752.
13 Fayette County Rolls Office Patent Book, No. 15, p. 97. (MSS.)
14 *Alden's Appeal Record,* Coleman vs. Brooke, p. 139.
15 Fayette County Tax Lists, 1789-1800; Westmoreland County Tax Lists, 1789-1800;
16 *Cazenove Journal,* p. 36.
17 J. Doddridge, *Notes on the Settlement and Indian Wars of the Western Parts of Virginia and Pennsylvania,* p. 108.
18 *Ibid.,* pp. 137-138; *Cazenove Journal,* p. 84.
19 Pennsylvania Furnace and Forge Ledgers. (Historical Society of Pennsylvania).
20 Potts MSS., XXIII, Pine Forge, 1748 (1748-1757), Day Book, pp. 45, 52, 88 ff. *passim; Ibid.,* Coventry, III, 1734 (1734-1741), pp. 19, 28, 29, 54; also *Pennsylvania-German* (1907), VIII, 126 ff.
21 Henry Drinker to Richard Blackledge, 10 mo. 4, 1786, Drinker Letter Book, 1786-1790, p. 81.
22 Peter Kalm, *Travels into North America* (1770 edition), I, 92-94.
23 Israel Acrelius, *History of New Sweden,* p. 145.
24 Peter Kalm, *Travels into North America,* 1. 93.
25 Before the Revolution several furnaces in parts of southeastern Pennsylvania had been forced to suspend operations because of lack of wood. See S. G. Hermelin, *Report About the Mines in the United States of America, 1783,* p. 74
26 Pennsylvania Furnace and Forge Ledgers.
27 *Cazenove Journal,* p. 33.
28 *American Husbandry,* I, 171; Peter Kalm, *Travels into North America,* I, 144-145.
29 *Cazenove Journal,* p. 33.
30 *Columbian Magazine* (1787), I, 34-39.
31 *Universal Asylum and Columbian Magazine* (1791), II, 27-31.
32 Pennsylvania Furnace and Forge Ledgers.
33 *Pennsylvania Gazette,* September 16, 1742; Thomas Cooper, *Some Information Respecting America,* p. 145; Cramer's *Pittsburgh Almanack,* 1812.
34 Pennsylvania Furnace and Forge Ledgers.
35 Henry Wansey, *Excursion to the United States,* pp. 152 *Pennsylvania Packet and Daily Advertiser,* July 8, 1788.
36 Henry Wansey, *Excursion to the United States,* pp. 152-153.
37 Duke de la Rochefoucauld Liancourt, *Voyage dans les Etats-Unis d'Amerique,* I. 3 ff.
38 Potts MSS., B XII, Pottsgrove, 1755 (1755-1765), p. 59. *Ibid.,* V, Pine Forge, 1732 (1732-1743), p. 175. Most of the ledgers contain entries of payments made to schoolmasters.
39 *Ibid.,* XXIII. Pine Forge, 1748 (1748-1757), p. 75; *Ibid.,* XV. Pottsgrove, 1758 (1758-1769), pp. 24, 75, 160, 171, 192, 234, 260.
40 A. S. Bolles, *Pennsylvania, Province and State,* II. 435-436.
41 *Ibid.,* II, 436.
42 C. F. Huch, "Henry William Stiegel," *Pennsylvania-German* (1908), IX, 72-73.
43 Mrs. T. P. James, *Memorial of Thomas Potts, Jr.* p. 387.

FOOTNOTES—Continued

[44] He wrote in part: "Next day I set off to my appointment at Potts [Warwick] Furnace, which for wickedness was next door to hell. Here they swore that they would shoot me. Mrs. Grace, hearing of their threats, and being herself unwell and not able to attend, sent a person to moderate the furnacemen and colliers." Benjamin Abbott, *Experience and Gospel Labors*, pp. 80-81.

[45] John Cuthbertson, Diary.

[46] It was computed that more than 200,000 gallons of rum alone were brought into the colony in 1728. Of this amount, only 11,400 gallons were re-exported. **More than £25,000 worth of rum was drunk in Pennsylvania during that year.** *Pennsylvania Gazette*, 11 mo. 7, 1728-29. For the latter part of the century, see J. D. Schoepf, *Travels in the Confederation*, 1783-1784, I, 363; F. A. Michaux, *Travels to the West of the Alleghany Mountains* (second edition), pp. 39-40. Before the close of the century, the manufacture of whiskey had become very profitable in the western country. Tench Coxe, *View of the United States*, pp. 51, 52.

[47] *Pennsylvania Packet and Daily Advertiser*, July 12, 1788, etc.

[48] The ledgers of the furnaces and forges contain a large number of entries of liquor, especially rum, bought by the workmen.

[49] *Pennsylvania Statutes at Large*, IV, c. 293.

[50] *Ibid.*, IV, c. 344.

[51] *Ibid.*, IV, c. 293.

[52] F. A. Michaux, *Travels to the West of the Alleghany Mountains*, pp. 39-40.

[53] In Pennsylvania, the term *borough* applies to incorporated units below the rank of cities. Before the Revolution, only a few boroughs were chartered: Philadelphia soon after 1684 (*Colonial Records of Pennsylvania*, I, 117); Germantown in 1691 (*Pennsylvania Archives*, First Series, I, 111-115); Chester in 1701 (Samuel Hazard, *United States Commercial and Statistical Register*, III, 264-265); Bristol in 1720 (*Ibid.*, III, 312-314); Wilmington, which is now in Delaware, in 1739 (*Ordinances of Wilmington*, pp. 145-150), and Lancaster in 1742 (Samuel Hazard, *United States Commercial and Statistical Register*, III, 397-398).
From the Revolution to 1800 the State legislature granted charters to the following boroughs:

Lancaster	1777	(*Pennsylvania Statutes at Large*, IX, c. 759.)
Carlisle	1782	(*Ibid.*, XI, c. 1029.)
Reading	1783	(*Ibid.*, XI, c. 1031.)
Bristol	1787	(*Ibid.*, XII, c. 1182.)
York	1787	(*Ibid.*, XII, c. 1315.)
Easton	1789	(*Ibid.*, XIII, c. 1438.)
Harrisburg	1791	(*Ibid.*, XIV, c. 1570.)
Pittsburgh	1794	(*Ibid.*, XV, c. 1771.)
Chester	1795	(*Ibid.*, XV, c. 1806.)
Bedford	1795	(*Ibid.*, XV, c. 1811.)
Huntingdon	1796	(*Ibid.*, XV, c. 1892.)
Uniontown	1796	(*Ibid.*, XV, c. 1910.)
Sunbury	1797	(*Ibid.*, XV, c. 1938.)
Greensburg	1799	(*Ibid.*, XVI, c. 2016.)
West Chester	1799	(*Ibid.*, XVI, c. 2044.)
Lebanon	1799	(*Ibid.*, XVI, c. 2045.)
Frankford	1800	(*Ibid.*, XVI, c. 2123.)

[54] W. P. Holcomb, *Pennsylvania Boroughs*, Johns Hopkins University Studies (1886), IV, 41.

[55] *Pennsylvania Statutes at Large*, XV, c. 1904.

[56] *Laws of Pennsylvania*, 1816, c. 4148.

[57] J. F. Sachse, "Wayside Inns on the Lancaster Roadside," Pennsylvania-German Society, *Proceedings* (1912), XXI, Part XXIII, pp. 5 ff. Mrs. A. M. Earle, *Stage Coach and Tavern Days*, pp. 143 ff. passim.

[58] H. M. Chapin, *Early American Signboards*, pp. 1 ff.

[59] *Ibid.*, pp. 8-9, 11.

[60] John Omwake, *The Conestoga Six-Horse Bell Teams of Eastern Pennsylvania*, pp. 21 ff.

[61] William Douglass, *British Settlements in North America*, II, 333. A. Burnaby, *Travels through the Middle Settlements*, 1759-1760 (1775 edition), p. 88.

[62] Potts MSS.: Colebrookdale Ledgers. The books of all other iron works show the same high rates for transporting iron. Potts MSS. B II, Coventry, 1728 (1727-1734), p. 80, etc.

[63] I. D. Rupp, *History of Dauphin, Cumberland, etc. Counties*, p. 376.

[64] Sherman Day, *Historical Collections of Pennsylvania*, pp. 354-355.

[65] A furnace and forge had been built in the Ohio country before 1799. *Pittsburgh Gazette*, February 2, 1799.

[66] The ironworks situated near the Schuylkill River sent part of their iron to Philadelphia in small boats. The records show that these boats carried from ten to fifteen tons. The cost of transportation was but slightly lower than by road. Potts MSS. LXX, Pine Forge, 1774 (1774-1781), pp. 55, 56, 114; *Ibid.*, LXXI, Pottsgrove, 1772 (1772-1789), pp. 272, 273; *Ibid.*, B II, Coventry, 1728 (1727-1734), p. 56. A memorandum of goods sent by boat down the Schuylkill River for the Philadelphia market from December 20, 1800 to June 20, 1801, includes 20,000 barrels of flour, 600 barrels of bread, 110

FOOTNOTES—Concluded

tons of iron, as well as great quantities of country products. *Pittsburgh Gazette*, July 10, 1801.

67 J. L. Ringwalt, *Development of Transportation Systems in the United States*, p. 13; Stewart Pearce, *Annals of Luzerne County*, pp. 455-456; Israel Acrelius, *History of New Sweden*, p. 165.

68 George Washington to President of Congress. December 8, 1776, J. Sparks, (Ed.) *Writings of George Washington*, IV, 206-207.

69 Boats of from twelve to fifteen tons burden could navigate the Juniata to within a mile or two of Bedford. It took four men to push such a boat up the stream. J. D. Schoepf. *Travels in the Confederation*, 1783-1784, I. 229.

70 Israel Acrelius, *History of New Sweden*, p. 165. See also A. C. Bining, *British Regulation of the Colonial Iron Industry*, Chapter V and Appendix.

71 Pennsylvania Furnace and Forge Ledgers.

72 As early as 1769 attempts were made to improve the inland navigation of Pennsylvania. American Philosophical Society *Transactions*, I. 293-300. It was not until the last decade of the century, however, that internal improvements were really begun. For a list of proposed improvements of roads and waters, see Thomas Cooper, *Some Information Respecting America*, pp. 39-47.

73 Emanuel Swedenborg, *Regnum Subterraneum sive Minerale de Ferro*, Part I, Section 13.

74 Peter Kalm, *Travels into North America*, I. 167.

75 Israel Acrelius, *History of New Sweden*, pp. 164 ff.

76 J. D. Schoepf, *Travels in the Confederation*, 1783-1784, I, 197-216, passim, II. 4-8.

77 F. A. Michaux, *Travels to the West of the Alleghany Mountains*, p. 48.

78 *Extracts from the Journal of Elizabeth Drinker*, p. 20.

79 Jacob Hiltzheimer, *Diary*, p. 36. Joseph Galloway was the owner of the Durham Iron Works at this time and had leased them to George Taylor.

80 George Washington to General A. Wayne, September 19, 1777, Wayne MSS., IV. Document 7.

81 W. S. Baker, *Itinerary of George Washington*, pp. 90-91.

82 "Journals of Captain John Montresor," New York Historical Society *Collections*, XIV. 457.

83 Lebanon County Historical Society, *Publications*, III, No. 1, 33-50.

84 R. P. Nevin, *Le Trois Rois*, pp. 57-58. There seems to be some grounds for the origin of this tragic legend. Marmie, who was impulsive and temperamental, lost all his wealth because of his connections with the changing French governments during the years following the French Revolution. No evidence, however, has been found to show that he took his own life.

85 It has been pointed out that wages in this country were paid to the iron workers largely "in kind."

86 *Pennsylvania Archives*, First Series, II, 57; *Pennsylvania Colonial Records*, IX, 635.

87 *Pennsylvania Archives*, Third Series, XI, 634, 752.

CHAPTER III

The establishment of the industry

THE purpose of this chapter is to sketch the geographical distribution of the ironworks of Pennsylvania during the eighteenth century and to point out the extent to which certain regions became industrialized, even during this early period. While the industry, as it originated, developed and spread, was quite scattered in its nature, there were forces at work that tended to concentrate it in certain regions. The trend can be noted from the first part of the century when iron manufacture developed in southeastern Pennsylvania, throughout the period to the end of the century, when, following the moving frontier westward, certain regions in the Monongahela country became to some degree industrialized. At least, in certain areas there can be traced in outline, industrial development that foreshadowed a new era that was not far distant.

Several factors determined the locations of the ironworks. An adequate supply of ore, an abundance of wood, sufficient water power, convenience of transportation and markets even though local, were essential for the successful operation of these early industrial plants. Of these requisites, the first was perhaps the most important. It was because of the discovery of good ores that works were built on plantations, often in remote regions, at times without regard to difficulties of transportation between the ironworks and the outside world. The forests and streams, important and necessary to iron manufacture during this period were considered lesser factors because of their abundance. The markets usually local at first, gradually broadened as population increased and settlement expanded.

The Schuylkill Valley

The earliest Pennsylvania ironworks were built along the

tributaries of the Schuylkill River. In the years following the erection of a bloomery forge in 1716 by Thomas Rutter[1] in the stillness of the forests of the Manatawny region, about forty miles from Philadelphia, many ironworks were set up in rapid succession. The period of inertia and delay was past and many plans were made. Colebrookdale Furnace, named after the famous English furnace of Abraham Darby in Shropshire, was the pioneer blast furnace of the colony. Erected on Iron Stone Creek, a branch of the Manatawny, by a company headed by Thomas Rutter,[2] this furnace, together with a forge, remained in operation until just before the

Ruins of Hopewell Furnace

Revolution. In the decades that followed the erection of Colebrookdale Iron Works, the Manatawny region became the scene of industry and Berks county for a time attained the industrial leadership of America.

Another region which developed industrially at the same time as the Manatawny district was that of Coventry, along the banks of French Creek. This winding creek furnished the power for many industries as time went on. About the same time that Thomas Rutter was erecting his bloomery forge, Samuel Nutt, Sr., who in the early documents is described as a weaver, began clearing the dense forests at Coventry for another bloomery, which was the beginning of the famous Coventry Iron Works. The fires were lighted in 1718.[3] In partnership with William Branson and Mordecai Lincoln, within the next few years refinery forges, and a blast furnace had been added to the works.[4] In 1732, Samuel Nutt, Sr. decided to experiment in the making of steel, and a steel furnace, the first in the province, was built.[5] It was in this region that Reading and Warwick furnaces — frequently mentioned by travelers as rivaling English furnaces as to size and production of iron before the Revolution—were established.[6] Among the important works of this period were the Oley Iron Works, operated by the ironmaster, Daniel Udree. It was in the Schuylkill Valley that the Potts, Rutter, Nutt, and other families held large interests in industrial enterprises.[7]

In the same region the Bird family began their many enterprises. The first was Hopewell Forge, built in 1744.[8] Before William Bird died in 1761, he had established three forges on Hay Creek,[9] and also Roxborough Furnace (later called Berkshire) in Heidelberg township.[10] After the death of his father, the ambitious Mark Bird decided to expand his interests. He built Hopewell Furnace in 1770 on French Creek.[11] Before the outbreak of the Revolution, he had added a slitting mill and nailery to his works,[12] and not long afterwards began to produce steel. The financial difficulties and the failure of Mark Bird will be discussed in another chapter. In 1788, his various ironworks were sold separately.[13]

It was during this period that Mount Joy Forge (Valley Forge) made famous by Washington and the sufferings of his army during the bleak winter of 1777-1778, was built on Valley Creek.[14] For more than a quarter of a century it produced iron to meet a local demand until it was destroyed by the British in 1777. For many years thereafter it remained in ruins. At the time, it belonged to Joseph Potts and William Dewees. The works were finally re-built on a much larger scale and Isaac Potts offered them for sale in 1790. They were known then as the Great Valley Works.[15]

During the Revolution, the works of the Schuylkill Valley were exceedingly active, and played a very important part in the struggle. After the Peace of Paris in 1783, the activity in building new ironworks in spite of the depression years during the period of Confederation increased.[16] Not only were blast furnaces such as Sally Ann, Mount Pleasant, Dale, Joanna, Mary Ann, District and others built, as well as many forges on the various tributaries of the Schuylkill, but slitting mills such as the Vincent mill and the Cheltenham works were also established.

As iron manufacture increased, several of the older furnaces were abandoned. As in other regions, when a locality was depleted of wood, works were at times transferred to other districts or were abandoned. In 1783, Hermelin wrote:

Ruins of Sally Ann Furnace

The abandoned ironworks are: The Reading blast furnace in Chester county, completely abandoned because of lack of timber and for other reasons; the Collebrookdale [Colebrookdale] blast furnace in Chester [Berks] county, abandoned partly because their own forests have been used up by coaling, and partly because the ore has become too expensive. The Moyberry [Maybury] blast furnace . . . is abandoned partly because of [lack of] timber, and partly because the ore must be transported . . . 10 miles.[17]

Competition with the newer furnaces had forced these pioneer furnaces to suspend.

Before the end of the century, much industrial progress had been made in the Schuylkill Valley. The charcoal-iron industry had been established and the ironworks were turning out pig iron and bar iron for local needs and for distant markets. Bar iron was being manufactured by hundreds of blacksmiths into tools and implements indispensable to a civilization that was becoming industrial conscious. The manufacture of hammered sheet iron, slit iron, nails, and the many iron articles needed in ship-building was increasing. In spite of many obstacles and difficulties in this agrarian age, a new era of industrialism was beginning to dawn.

The Delaware Valley

The earliest enterprises in ironmaking in the picturesque Valley of the Delaware were attempted with but slight success in a part of the region that is now the state of Delaware. Almost at the same time that Thomas Rutter and Samuel Nutt were building their forges in the Schuylkill Valley, John Ball was planning a bloomery on White Clay Creek in New Castle county.[18] This works, known as the St. James' Church Bloomery met with indifferent success. Other attempts, made in this section, owing to a scarcity of good ores and difficulties in financing, resulted in failure. The enthusiastic Sir William Keith, the industrialist, John England of the Principio Company, Maryland, as well as the eight Pennsylvania ironmasters including Thomas Rutter, who built the Abbington Iron Works on Christiana river, met with disaster.

One of the most important and famous works in the Delaware Valley was the Durham Iron Works, founded in 1727 by a company of twelve men, including some of the leaders in the colony at the time.[19] The career of this works, which during most of its existence included a blast furnace and three forges, came to an end in 1789. The site, however, was used for later furnaces. Among the prominent owners of the eighteenth century Durham Iron Works were James Logan, Anthony Morris, William Allen, Joseph Turner, Joseph Galloway, and George Taylor.

Many small individual forges were built along the Delaware during this period. Crum Creek Forge was the first enterprise of Peter Dicks, who later extended his interests into York county.[20] Springton Forge on the Brandywine, Solebury Forge near Coryell's Ferry,[21] and others began to dot the Pennsylvania side of the valley. Before 1739, on Chester creek John Taylor began his industrial experiments that led to the development of the Sarum Iron Works.[22] Here in 1746, the first slitting mill in the province was established.[23]

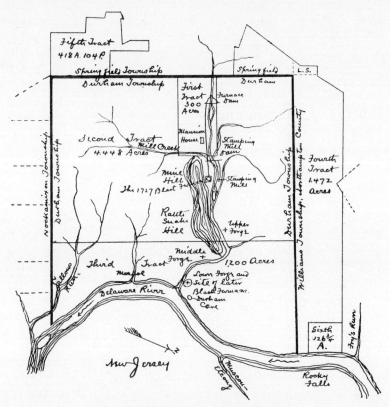

Plan of Durham Iron Works Property

After the Revolution some ambitious projects were undertaken. The Delaware Works belonged to Mark Bird and James Wilson. After the failure of Mark Bird, James Wilson carried on for a time this enterprise, which consisted of slitting mills, a forge, grist mills and saw mills.[24] Hermelin remarked about this project in 1783:

A similar plant [has been established] this year at the lower fall of the Delaware River, opposite Trenton, with four water wheels for iron strips and hoop-iron, partly in order to endeavor to manufacture round bolt-iron for shipbuilding out of square blocks of iron under concave rollers.[25]

Among the smaller works erected during this period, were Mary Ann Forge in 1785, on the north branch of the Brandywine, Rockdale Forge in Aston township by Abraham Pennell, Hibernia Forge in 1793 on West Brandywine Creek, and Federal Slitting Mill on Buck Run by Isaac Pennock in 1795. This latter works became the Rokeby Rolling Mill.

Even the Quaker city, Philadelphia, was becoming industrialized before the century closed. Ironworks of various kinds were built within the limits of the capital city. Stephen Paschall's Steel Furnace was producing steel in 1747. William Branson also engaged in the production of blister steel before the middle of the century.[26] Daniel Offley's Anchor Forge was near the water front, where "through the thick sulphurous smoke, aided by the glare of light from the forge, might be seen Daniel Offley, directing the strokes of a dozen hammermen, striking with sledges on a welding heat produced on an immense unfinished anchor, swinging from the forge to the anvil by a ponderous crane, he at the same time keeping his piercing iron voice above the din of the iron sound."[27] In spite of the act of 1750 prohibiting the erection of steel furnaces, Whitehead Humphreys built such works about 1762 in the center of the city.[28] Nancarrow and Matlack later rebuilt their steel furnace not very far away.[29] Caleb Foulke owned a furnace in the same neighborhood in 1794.[30] Many air furnaces, tin plate factories, and nail works were established in Philadelphia from the time of the Revolution to the close of the century.[31] Writing of this region at the close of the Revolution Hermelin expressed his impressions as follows:

> Hand-smithing of all kinds is done to a great extent in the country districts and in the many small towns. Such manufacture takes place in towns of the Moravian Brothers, Bethlehem, Nazareth and Letiz [Lititz]. Scythes, famous [for quality], are made at Derby [Darby]. Steel and wire mills for card manufacture are found in Philadelphia and [in] Wilmington, in the

state of Delaware. At Philadelphia and the nearby Kensington there are several smithies for manufacture and repair of anchors and the manufacture of bolt-iron and other iron work for ship-building, for which [manufacture] hard coal is brought by ships from Richmond in Virginia, where I observed good coal strata in considerable quantities. The smiths in Pennsylvania work independently, but in the Southern States most of them are employed at plantations owned by others.[32]

THE SUSQUEHANNA VALLEY

In the broad Susquehanna Valley as in the other valleys, while the industry was scattered, there were centrifugal tendencies at work that seemed to concentrate the ironworks in a few centers. In the regions of Lancaster, Lebanon, York and Cumberland, on many of the tributaries that carried their waters into the slow-flowing Susquehanna River, ironworks were built.

The pioneer works in the Lancaster region was Kurtz' Iron Works, a bloomery, built about 1726 on Octorara Creek.[33] Little is known of the activities of the forge. The iron industry of the Susquehanna Valley did not make progress until after Peter Grubb built Cornwall Furnace in the nearby Lebanon region. The first Windsor Forge was built by William Branson in 1743, between the Forest Mountains and the Welsh Mountains.[34] A second forge was added by Lynford Lardner, who carried on the works with Samuel Flower and Richard Hockley.[35] In 1773, a half interest in the property came into the hands of David Jenkins, and in 1775, he owned the entire property. The Jenkins family operated the forges for a long period of time until they were abandoned in 1850.

The Lancaster region was the scene of Henry William Stiegel's unwise speculations. John Jacob Huber had a furnace on a tributary of Conestoga Creek about 1750.[36] In 1757, Stiegel bought the furnace and rebuilt it, in partnership with John Barr and Alexander and Charles Stedman. The story of Stiegel's meteoric rise and pathetic fall in the iron and glass industries will be told in a future chapter. Among other enterprises in this region was Martic Iron Works, built on a branch of Pequea Creek in 1754 by Thomas and William Smith. The chain of title to the property discloses a long line of owners. Speedwell Forge, built by James Old in partnership with David Caldwell and Poole Forge were among the early ironworks in this section of the province.

During the last two decades of the century much industrial progress was made. Near a ravine on Big Chiquisalunga Creek, Peter Grubb, Jr., established Mount Hope Furnace in 1785. Cyrus Jacobs built Spring Grove Forge on Conestoga Creek, about three miles west of Poole Forge in 1793. Among the smaller establishments before the end of the century, in the Octorara region were Pine Grove Forge, Sadsbury Forge and Duquesne Forge.[37]

The Lebanon region became noted early because of the discovery and development of the Cornwall ore deposits. Before the Lake Superior ores were utilized, and even down to recent times the Cornwall mines produced more iron ore than any other single mining property in the United States. The importance of these mines to the industrial development of the United States may be seen in the fact that more than 20,000,-000 tons had been produced from Cornwall by 1907.[38]

The history of the Lebanon mines began when the proprietors, John, Thomas and Richard Penn conveyed the tract containing the Cornwall ore hills to Joseph Turner of Philadelphia in 1732, who assigned them to William Allen in 1734, who in turn, a few years later, transferred them to Peter Grubb. Cornwall Furnace, built in 1742, succeeded a bloomery established by Grubb and a beginning was made in the opening of the great ore deposits of the area. Forges were erected as part of the Cornwall Works. Grubb died in 1754 and his sons, Curtis and Peter Grubb inherited the estate. In time, these works, together with Speedwell Forge, and other properties were owned chiefly by the ironmaster, Robert Coleman. The reservation of the right for a sufficient quantity of ore for one furnace by Peter Grubb, Jr., the son of Curtis Grubb, resulted in a famous lawsuit in the nineteenth century.[39] The mines at Cornwall, still in operation are now owned by the Bethlehem Steel Corporation. Among the early works in this region in addition to those already mentioned was Quittapahilla Forge. Colebrook (Mount Joy) Furnace on Conewago

Creek in 1791, was built by Robert Coleman and ran until just prior to the Civil War.

West of the Susquehanna, in York county, another group of works sprang up. Peter Dicks who had started his career in the Delaware Valley, discovered iron and built a bloomery before 1756.[40] In 1763, the estate was sold to George Ross, George Stevenson and William Thompson,[41] who a year before began building Mary Ann Furnace on Furnace Creek.[42] They also built a refinery forge called Spring Forge.[43] As was the case with many pioneer ironworks these properties changed

Recent View of Cornwall Mines

hands frequently. Mary Ann Furnace ran until the beginning of the nineteenth century.[44] Codorus Iron Works—a blast furnace and forge—were erected by William Bennett on Codorus Creek. It was sold by the sheriff in 1771, and soon came into possession of James Smith, who after heavy losses sold it during the Revolution.[45] The history of this works was a checkered one during this period. Like many other furnaces, both Mary Ann Furnace and Codorus Iron Works cast cannon balls during the Revolution.

Still another region of the Susquehanna Valley that produced iron quite early in its history was the Cumberland Valley. The Carlisle Iron Works (Boiling Springs) were established in 1762 by John Rigby and Company,[46] although two years later, the company consisted of Robert Thornburg, Joseph Armstrong, Samuel and John Morris, and Francis Sanderson.[47] Michael Ege became the sole owner in 1792. On the top of the South Mountain, by the rushing waters of Mountain Creek, midway between Carlisle and Gettysburg, Pine Grove Furnace was built in 1764 by George Stevenson, Robert Thornburg and John Arthur.[48] By the beginning of

Ruins of Pine Grove Furnace

the nineteenth century, this furnace also became the property of Michael Ege.[49] Mount Holly Iron Works was another of George Stevenson's enterprises, and was built at Mount Holly Springs.[50] Just outside Carlisle, an armory which had been established in the sixties was exceedingly active during the Revolution. Muskets, locks, sabres, as well as cannon forged from iron gads and hoops, soldered together, were made during the war. With the close of the war, the works were removed.

After the Revolution a number of ironworks were built in this region, the most important of which was the Cumberland Furnace, which ran for sixty years. On the whole the ironworks of this region were quite successful and as in other regions, increase of population demanding iron for so many uses resulted in a broader home market.

In the western part of Cumberland county—the region that became Franklin county—a bloomery forge was built by William, Benjamin and George Chambers in 1783, not far from Fort Loudon. This small plant, which supplied many frontiersmen with needed iron as they marched westward, was the nucleus of the Mount Pleasant Iron Works. Not many miles away, the Loudon Iron Works were built about 1790 by James Chambers.[51]

THE JUNIATA REGION

Soon after the French and Indian War had been brought to a successful conclusion, and Great Britain's march of empire westward was now no longer contested by the French, many from the frontier turned their eyes to the stretches of choice lands that lay in the west. Most of those who traveled into the Juniata regions were interested in lands and agriculture. However, as early as 1767 a company called the Juniata Iron Company was organized by a group led by Joseph Jacobs. Surveys were made and iron deposits were found in the years that followed, but as the Revolution came on the plans to utilize the ores of this section were dropped.

After the Revolution was over and as the westward movement got under way again many industrial projects were undertaken. It was not until 1785, however, that Bedford Furnace and Forge were erected by George Ashman and Company.[52] His partners included Charles Ridgely, Thomas Cromwell and Tempest Tucker. Among other works established in this region were Licking Creek Forge, erected in 1791 by Thomas Beale and his associates on Licking Creek, a branch of Tuscarora Creek, Freedom Forge in 1795 by William Brown in Mifflin county, and Barree or Dorsey Forge built by Greensburg Dorsey about the same time.[53] Huntingdon Furnace [Anshutz Furnace] on Warrior's Mark Run was established in 1796 by Mordecai Massey and John Gloninger, and was managed by George Anshutz. Hope Furnace was established in 1797 by William Lewis, after whom Lewistown was named. Thomas Evans became associated with him in operation of the furnace. Juniata Forge was built by Dr. Peter Shoenberger who laid out Petersburg, and Spruce Creek Forge by Phineas Massey before the close of the century. The latter forge soon passed into the hands of John Gloninger and Company. Caledonia Forge, near Bedford Springs, and William Lane's Hopewell Forge in Bedford county were in operation before 1800. By this time a steel furnace was built by William McDermett at Caledonia near Bedford.[54]

Another part of this region—that which became Centre county in 1800—witnessed the rise of industry during the last decade of the century. Many Revolutionary veterans engaged in these enterprises. In 1791 Colonel John Patton and Colonel Samuel Miles built Centre Furnace, which had a long and successful career. General Philip Benner, who had learned ironmaking at Coventry Forge, and who had fought in the Revolution with Anthony Wayne erected Rock Forge and Slitting Mill in 1793. Other works were built—Spring Creek Forge in 1795 by Daniel Turner, Harmony Forge the same year by Colonel Samuel Miles, Colonel James Dunlop and John Dunlop. The latter built Bellefonte Forge in 1795 and just before the turn of the century Logan Furnace.[55]

In this industrial section the ironmasters struck out on a new trade with Pittsburgh, which was becoming a gateway to the west. Bar iron, slit iron and nails were first sent on mule back over almost impassable roads, salt especially being brought back in return. As roads were laid out, the trade between the Juniata regions and Pittsburgh increased. The development of markets down the Juniata and the Susquehanna Rivers as well as in the west belongs to the period after 1800 and will be discussed in another volume. It is evident, however, that a start had been made in the production of "Juniata iron" which in the nineteenth century was to be known not only in America, but in Europe.

OVER THE ALLEGHENIES

The origin of the great iron and steel industry that now centers in western Pennsylvania can be traced to the period when the Monongahela country was a part of the frontier; for it was not long after the vanguard of pioneers had pushed into this region that iron manufacture began. Before the outbreak of the American Revolution relatively few settlers ventured west of the mountains. This was partly due to Great

Britain's policy of attempting to restrict settlement to the fringe of territory along the Atlantic coast. Migration was also retarded because of the large tracts of unoccupied lands still remaining just east of the mountains. During the struggle with England and even earlier, stories of green valleys, clear streams, and abundant game, told by explorers and traders, lured a few of the sturdy sons of Pennsylvania and Virginia to the Alleghenies, even as far as the beautiful Ohio River. Here some of them settled, building stockades as a means of protection against the untamed Indians and clearing green woodlands in order to plant fields of golden grain. After independence had been wrested from the mother country, emigrants in increasing numbers sought the newly-opened West, and the Monongahela region became one of the great highways over which thousands and thousands of stalwart pioneers passed on their arduous and difficult trek westward—pioneers who looked forward to making homes for themselves in the fertile wilderness, staking their strength, courage and ambitions against isolation, privation and hardship.

Many factors played a part in bringing settlers to the western country after 1783. The removal of British restrictions, the five years of dismal business depression following the treaty of peace, the failure of crops in Virginia, the encouragement given by crafty land speculators, the land hunger of veterans of Washington's armies, the passage of the Northwest Ordinance, Anthony Wayne's brilliant victory over the Indians at Fallen Timbers, and important changes in the land laws, all encouraged eager home-seekers and discontented citizens from the East to migrate westward. Before the close of the eighteenth century thousands had traveled over the mountains to the Monongahela River and from there on boat and raft to the Ohio River. Many, on reaching western Pennsylvania, went no farther but took up land adjacent to the waterways of the region, while others traveled on in their search for economic security.

Even on the quiet frontier, far away from the rankling cries of civilization, iron was required for many uses. Horses had to be shod and wagon tires often needed repairs; nails, hinges, and bolts were essential for buildings; and strong axes were necessary for clearing the dense forests. Among the earliest pioneers were a few blacksmiths, who took with them on their journey westward small quantities of hammered bar iron from eastern forges, which they shaped to meet frontier needs. From the earliest days of settlement bar iron was sent to the frontier. As might be expected, however, there was always a scarcity of iron in all forms, and hastily built cabins and houses were often put together with wooden pins and pegs. Many stories have been told of the burning of deserted structures for the purpose of procuring the iron nails they contained. Even the temporarily abandoned Fort McIntosh was almost entirely destroyed by westward migrators, who drew its nails and secured from it other material to aid them in building homes.

Because of the great need for iron in the new country, it was not long before enterprising men seized the opportunity to engage in iron manufacture, and ironworks of different

Bloomery Forge

kinds arose amidst the shadows of deep green forest glades, where Indians still lingered. The flames from the stacks of blast furnaces, forced upward by the rhythmic blast, which at night cast a glare on the sky like a brilliant display of northern lights, the scintillating sparks scattered in the forges by giant hammers striking the red-hot iron bars, the ringing sound of the anvils of many blacksmith shops where iron was shaped into needed articles, and the splashing of water over large water wheels, which furnished the only power for these early manufacturing plants, soon gave many parts of the agricultural frontier a glowing tinge of industrial activity.

It was natural that the Fayette county region should take the lead in establishing iron manufacture in western Pennsylvania because it was along the populous line of travel westward and possessed all the natural advantages necessary for the production and manufacture of iron. Rich ores were found in many places, wood for fuel was plentiful, limestone could be secured, and water power was abundant. A list of the first works built in this section of the frontier clearly illustrates the rapidity with which iron manufacture took hold. Among the pioneer ironworks established during the last decade of the eighteenth century in this region were the Alliance (Jacob's Creek) Iron Works, Hayden's Bloomery, Fairfield Furnace and Forge, Laurel Furnace, Little Falls Bloomery, Mary Ann (later Fairview) Furnace, Mount Vernon Furnace and Forge, Pears' Bloomery and Slitting Mill, Pine Grove Forge, Redstone Furnace, Spring Hill Furnace and Sylvan (Oliphant's) Forge, Union Furnace and Forge, and Youghiogheny (Lamb's) Forge.[56] In three other sections of western Pennsylvania ironworks were built before 1800: the ill-fated Greene Furnace in Greene county,[57] Westmoreland Furnace and Forge in the thickly wooded Ligonier Valley,[58] and the short-lived furnace of George Anshutz at Shadyside, Pittsburgh. Thus before the dawn of the nineteenth century—a century that was to bring so many industrial changes—iron manufacture had been well planted and had taken root in western Pennsylvania.

FOOTNOTES

1 Jonathan Dickinson to James Logan, 1717, Dickinson Letter Book, 1713-1721; Minute Book 4, Board of Property, *Pennsylvania Archives*, Second Series, XIX, p. 651. *Philadelphia Weekly Mercury*, Nov. 1, 1720, *Pennsylvania Gazette*, March 5-13, 1729-30.

2 The chain of title may be found in Berks County Deed Book, B, I, 358-360; Land granted by David Powell to Thomas Rutter, Patent Book, A, V, 377: Book F, II, 206 ff.

3 Minute Book II, Board of Property, *Pennsylvania Archives*, Second Series, XIX, p. 620; *Ibid.*, 612; Samuel Nutt to Isaac Taylor, July 2, 1720, Taylor Papers, XIV, Document 2960. Penn's Lessees *vs.* Kahler, 1810, Penn's Lessees *vs.* Mack, 1810, in Montgomery Collection of MSS.; Readinger Zeitung, March 4, 1795.

4 B. Potts MSS. II, Coventry 1728 (1727-1734) pp. 125, 130, 138 *passim;* Sur-

vey Warrant, July 13, 1733, p. 101, B I, copy in Montgomery Collection of MSS.

5 B. Potts MSS., II. Coventry 1728 (1727-1734) pp. 214, 239 *Ibid.* III, Coventry 1730 (1730-1732) pp. 132, 178, 371.

6 Israel Acrelius, *History of New Sweden*, p. 168; S. G. Hermelin, *Report About the Mines in the United States, 1783*, p. 72.

7 For a list of the many ironworks built in Pennsylvania see *Appendix* A.

8 Wm. Bird, Account Book, 1742-1750.

9 New Pine Ledgers, October, 1756, New Pine Forge Cole Book. Petition of Mark Bird to Orphan Court, May 30, 1763. Orphans Court Record Berks County, 1752-1792, I, 165; Anna Krick, MSS.

10 New Pine Forge Ledger, W. Bird Ledger, p. 66. Roxborough Furnace Ledger, 1758-1760. Eventually the furnace became the property of George Ege (1790) Berkshire Books. Berks County Deed Book A II, p. 544. *Reading Weekly Advertiser*, March 8, 1800.

11 C. B. Montgomery Collection of MSS.

12 *Pennsylvania Packet*, March 26, 1788: Deed Book A, Vol. 8, pp. 402 ff: Potts MSS. LXX, Pine Forge, 1774 (1744-1781) p. 188, *Ibid.* LXXIII, Pine Forge 1781 (1781-82) pp. 10, 18.

13 *Pennsylvania Packet*, March 26, 1788. Berks County Deed Book A. Vol. 8, pp. 402 ff.; *Reading Weekly Advertiser*, May 27, 1791, Dec. 14, 1799, Aug. 29, 1801.

14 *Pennsylvania Gazette*, April 4. 1751; September 26, 1751; Philadelphia County Road Docket, III, 155, 167, 171.

15 *Pennsylvania Weekly Mercury*, October 13, 1790; *Pennsylvania Gazette*, November 3, 1790.

16 See *Appendix* A.

17 S. G. Hermelin, *Report About the Mines in the United States of America*, 1783, p. 74. Reading Furnace was re-built and put into operation again in 1792.

18 Minute Book, I, Pennsylvania Board of Property, *Pennsylvania Archives*, Second Series XIX, 714.

19 Philadelphia Deed Book, G, III. 240-245.

20 Israel Acrelius, *History of New Sweden*, p. 165.

21 *Pennsylvania Chronicle*, August 23-31, 1767.

22 William Allen to John Taylor, August 18, 1739. Extracts and Memoranda of the Private Papers of John Taylor, Folio, Book, A, Cope Collection, p. 177.

23 *Pennsylvania Archives*, First Series, II, 57.

24 Papers of James Wilson, VII, 21, 84.

25 S. G. Hermelin, *Report About the Mines in the United States of America*, 1783, pp. 74-75.

26 *Pennsylvania Archives*, First Series, II, 57.

27 J. F. Watson, *Annals of Philadelphia and Pennsylvania*, I, 430.

28 *Pennsylvania Chronicle*. December 7, 1767, October 22, 1770, January 21, 1771, November 18, 1771.

29 *Pennsylvania Packet*, August 4, 1787; *Pennsylvania Mercury*, August 10, 1787.

30 *Pennsylvania Packet*, January 22, 1795.

31 *Ibid.*, September 4, 1797; *Philadelphia Federal Gazette*, April 21, 1795.

32 S. G. Hermelin, *Report About the Iron Mines in the United States of America*, 1783, p. 75.

33 Sherman Day, *Historical Collections of Pennsylvania*, p. 388.

34 Land Patent Book, A, X, 557 ff.

35 Lancaster County Commissioner's Minute Book, I, 109.

36 Elizabeth Furnace, Waste Book, A.

37 Miscellaneous papers regarding Mount Hope Furnace and some of the other ironworks may be found in Furnace MSS. Pennsylvania Archives, Harrisburg.

38 J. M. Swank, *Progressive Pennsylvania*, p. 219.

39 *Alden's Appeal Record*, Coleman *vs.* Brooke.

40 Israel Acrelius, *History of New Sweden*, p. 166.

41 York County Deed Book, B, pp. 50 ff.

42 *Ibid.*, A, pp. 520, 609, 612.

43 *Pennsylvania Gazette*, September 3, 1777.

44 Mary Ann Furnace Day Book, 1773.

45 W. C. Carter and A. J. Glossbrenner, *History of York County*, Appendix, p. 10

46 Cumberland County Deed Book, A, II, 289 f.

47 *Ibid.*, A, II, 291 f.

FOOTNOTES—Concluded

48 Cumberland County Deed Book, F, I, 299 ff.

49 T. P. Ege, *History and Genealogy of the Ege Family,* pp. 92-94.

50 Cumberland County Deed Book, K, I, 136; L, I, 357 f.

51 S. G. Hermelin, *Report About the Mines in the United States of America, 1783,* pp. 45, 74.

52 Huntingdon County Tax Lists 1787. *Pittsburgh Gazette,* November 30, 1793.

53 Huntingdon County Tax Lists, 1790-1800; Mifflin County Tax Lists, 1790-1800.

54 Bedford County Tax Lists 1788-1800.

55 J. B. Linn MSS.; J. Thomas Mitchell MSS.; J. B. Linn, *History of Centre and Clinton Counties,* pp. 20-30.

56 Fayette County Road Docket, I, 51, 67, 68, 72, 89, 93, 98, 99, 101, 103, 120, 121, 143, 145, 158, 170, 174, 182, 201; Fayette County Deed Book A, p. 392; Fayette County Deed Book B, pp. 39, 319; Fayette County Deed Book C, II, 758-760, 896-898; Fayette County Deed Book C, III, 1076-1079, 1202; Fayette County Township Property Rolls, 1796-1800.

57 Greene County Deed Book I, p. 748-750, Greene County Deed Book II, 70.

58 Westmoreland County Deed Book II, 57; Westmoreland County Deed Book III, 54, 170.

CHAPTER IV

THE TECHNIQUE OF IRON MANUFACTURE

DURING the eighteenth century, all the materials necessary in the manufacture of iron were found in abundance in Pennsylvania. Iron ores, outcropping on the surface of the earth, in veins, beds, irregular deposits, and to a small extent taken from bogs and ponds, were smelted and worked. The vast primeval forests and woodlands, which had been the peaceful home of animals and birds for centuries, and through which Indians had roamed, provided the material for the fuel used in furnace and forge. Beds of limestone, laid down in past ages, suitable for fluxing the ores of their impurities, were abundant. The many streams and creeks furnished the water power necessary to drive the tilt and helve hammers, the shears, cutters and other machinery, as well as the bellows or blowing cylinders which furnished the blast for furnace and forge.

The element iron is widely distributed over the surface of the earth and makes up about 4.44 per cent (weight) of the earth's crustal rocks.[1] Since this metal is found in many forms of rock, clay and earth, only those masses which contain sufficient metal to justify smelting are used as ores. With the exception of meteors, iron is never found in its pure state, but always exists in combination with other elements. Thus the gangue—the mineral material associated with the iron— may include sulphur, phosphorus, manganese, alumina, and other substances.

Iron ores are classified as follows: magnetite, red hematite, brown hematite or limonite, and carbonate.[2] All these types were mined and smelted in Pennsylvania from the earliest days of the iron industry. Magnetite ores (Fe_3O_4) occur in the southeastern sections of the state in Adams, Berks, Bucks, Chester, Columbia, Dauphin, Delaware, Lancaster, Lebanon

Montgomery, Northampton, Philadelphia, and York counties.[3] They contain 74.4 per cent of iron and 25.6 per cent of oxygen when pure, and are usually hard, heavy, black substances which give a characteristic black streak on a porcelain test plate. These ores are magnetic, and while the gangue is more or less complex, it is generally siliceous.[4] The black iron sand, found in colonial days on the banks of the Delaware and other American rivers, belongs to this class.[5] This heavy, black oxide, which was really composed of minute fragments of metamorphic rock, was not smelted successfully in the furnaces, although ironmasters experimented with it.[6] The sand had its uses, however, being a necessity for colonial scribes who used it for writing sand in days before blotting paper was known. The most important magnetite bodies and the most productive iron ore mines in the country until the development of the Lake Superior regions have been those of the famous Cornwall mines, worked continuously for 200 years, down to the present. Most of the ores produced in the state even at the present time are taken from these mines.[7]

Red hematite ores (Fe_2O_3) may be found in Adams, Berks, Bucks, Chester, Dauphin, Delaware, Huntingdon, Lancaster, Lebanon, Lehigh, Northampton, Philadelphia, and York counties.[8] They contain 70 per cent of iron and 30 per cent of oxygen when pure. They vary in appearance and physical character from very dense, friable types of a lighter shade of red. The gangue is variable to a great extent. These ores give a streak from cherry red to reddish brown when tested. The bedded red hematites of Clinton age have been widely worked in the past, but in recent years, owing to the opening up of Lake Superior ore regions, there has been little opportunity for much development for ores of this type.[9]

Brown hematite ores ($Fe_2O_3xH_2O$), part of the rock formations in Bedford, Berks, Bucks, Centre, Chester, Delaware, Franklin, Huntingdon, Lancaster, Montgomery, Northampton, and York counties,[10] contain varying quantities of water. The purest have not more than 59.9 per cent of iron. Bog ores, found in bogs and swamps, which resulted in the early development of the industry in Massachusetts, New Jersey and other colonies, belong to this class. The color and physical characteristics of the brown hematite ores vary considerably, from brown to yellow, and from dense and hard to light and friable. They give a yellowish brown streak when tested; the content of iron varies greatly. Brown hematite ores were mined extensively in the early days of the colonial industry, but today the output in Pennsylvania is small.[11]

Carbonate ores ($FeO. CO_2$) occur in many of the counties of Western Pennsylvania as well as in Berks, Chester, Delaware, Lancaster, Lehigh, Lackawanna, Lycoming, Montgomery, Philadelphia, Schuylkill, and York counties.[12] They abound in the anthracite and bituminous coal measures. The purest of them contain 48.3 per cent of iron, but they are variable. The gangue is usually calcareous; the characteristic streak is gray. Being the poorest of all the types of iron ores, the carbonates are not mined to any extent in the United States today. In 1923, Pennsylvania, the only state to mine this type of ore, produced 3,516 tons. In 1924, Pennsylvania produced 3,005 tons, and Ohio, 244 tons.[13]

Sulphides of iron, found throughout the state, are not considered or classified as iron ores. They are not worked for the iron they contain on account of the difficulty and expense of getting rid of the sulphur. Nor will they be worked until other ores give out. The pyrites, commonly calls "fools' gold" partly because they deceived early gold seekers who mistook the almost worthless ores for gold, are the most abundant of the sulphides. They are pale brass yellow in color, hard, brittle, and easily reduced to powder.

It was natural that men were early attracted to engage in the manufacture of iron in colonial America, for ores were plentiful. Acrelius, the Swedish pastor at Christina during the middle of the century, was astonished at the amount of iron ore in Pennsylvania. He wrote that there was more in the country than the people could ever use. Ore was found even among the loose stones on farmlands. After rains, long streaks of iron color, denoting the presence of iron in soil or stones, could be seen along the public roads and highways.[14] Foreign travelers and visitors frequently expressed their astonishment at the abundance of iron scattered over the earth in so many places.[15]

Since the ores first used were on the surface or just below, little or no technical knowledge of mining was necessary. The chief tools required in order to mine the ores were bars and pick-axes. There was little boring, blasting or firing.[16] Thus the German traveler, Schoepf, wrote:

> Any knowledge of mining is superfluous here, where there is neither shaft nor gallery to be driven, all work being at the surface, or in great, wide trenches or pits.[17]

These trenches were rarely more than forty feet deep. When this depth was reached, new "mine holes," as they were called, were opened. The depth of the mines at Oley was twenty feet.[18] At Warwick and Reading they were no deeper. The greatest depths reached at Cornwall and Durham prior to 1750, did not exceed forty feet.[19] Schoepf, watching the mining operations at Cornwall, compared them to the methods used in digging out paving stones at quarries. For this reason, Hermelin, the Swedish observer of mines in the United States after the Revolution, refers to them as "ore quarries."[20]

There is one record, however, of an attempt to operate a shaft mine in Pennsylvania prior to the Revolution. This was near Daniel Udree's ironworks, situated snugly in a valley among the Oley hills, about ten miles from Reading. A gallery had been driven into the hill, twelve feet high, fifteen feet broad, and about three hundred feet long. Then a shaft sixty feet deep was sunk. The ore obtained was rich and very easily fluxed in the furnace, but many difficulties were encountered in mining operations. Being a shaft mine, the ore had to be blasted. The scarcity and expensiveness of powder during the Revolution and the difficulty of getting experienced miners to work in such a mine doomed it from the beginning, and when water flooded it continually, all work had to be given up. Hermelin, however, ascribes its abandonment chiefly to the high cost of operation.[21] After the Revolution, other attempts were made to work this mine and apparently other shaft mines appeared. In New Jersey there were several mines of this type.

The number of miners employed at each mine during the eighteenth century was not large. Three or four could supply all the ore needed for a single furnace. The six miners employed by Daniel Udree at Oley in 1783 produced more iron ore than the furnace could use.[22] In 1781 the Durham Iron works advertised for five experienced miners to begin work again there.[23] At Bird's mine in Berks county, two miners were able to dig enough ore to keep the furnace in constant operation.[24] Henry Drinker estimated that three miners were sufficient to supply ore for one blast furnace.[25]

The fuel used to smelt the ores throughout the eighteenth century was charcoal. The ironmaster in England who used charcoal in his furnaces during the greater part of this period could no longer depend entirely upon supplies produced from forests or woods, for most of these had been destroyed and laws had been passed by Parliament restricting the use of forest timber in smelting iron.[26] To secure the much needed timber, coppices were preserved and kept especially for the ironworks.[27] After a period of experimentation, coke displaced the use of charcoal in English furnaces and during the latter part of the century the English ironmaster was no longer dependent upon wood for his fuel. But in the New World virgin forests which seemed at the time limitless in extent were cut down to provide fuel for the never-satiated furnaces. The use of coke as a furnace fuel was not considered until after more than three decades of the nineteenth century had passed.

The charring of wood in charcoal making was a very familiar process to the inhabitants of Pennsylvania in the eighteenth century, for charcoal was used as fuel in ironworks, blacksmith forges, by silversmiths, coppersmiths, and in many other crafts and trades. Wood was charred, not only in the country districts, but often within the limits of boroughs. The smoking piles of wood in process of being charred, which looked very much like Indian wigwams on fire, gave forth a dark heavy smoke of disagreeable odor. It was for this reason that many boroughs in Pennsylvania passed ordinances prohibiting the manufacture of charcoal within the borough limits.

Charcoal has been known for more than two thousand years, and was used by ancient peoples including the Hebrews and the Romans. Pliny describes the Roman method of making charcoal in his *Natural History*,[28] a process which had not changed essentially even in the eighteenth century. The process of charring in open piles was used in America until well into the nineteenth century, and even down to recent times. As coke came into use in England, coking was done at first in open piles, but this method was superseded by the use of ovens beginning in 1763.[29] The application of scientific methods and the adoption of retorts are improvements of more recent years.

Charcoal, however, remained the sole fuel for blast furnaces in the United States until about 1840, when after several years of experimentation, anthracite and bituminous coal, and finally coke were successfully used. Long before this, coke was used in the furnaces of England. Throughout the seventeenth century attempts had been made, especially by Dud Dudley, to produce a coke that would smelt iron successfully. The erratic and litigious Dud Dudley had carried out many

Charcoal House, Elizabeth Furnace

experiments but in spite of his boasts was unsuccessful in showing the practicability of his work.[30] Abraham Darby, the Quaker ironmaster of Coalbrookdale was first to use coke successfully. This was in 1709 or 1713, and from about this time coke became the fuel at his furnace, first for casting pots and other utensils, and after 1750 for making pig iron that could be forged into bar iron.[31] Coke was not used, however, by other English ironmasters, although there were many experiments made, until after the midway point of the century had been passed. The establishment in 1760 of the famous works at Carron and the large furnace at Seaton, where coke was the fuel, advertised the new process, and after this time, it was rapidly adopted by ironmasters all over England.[32] In spite of this change which took place in England, American ironmasters continued to use charcoal until well into the nineteenth century. There were many reasons for this. The chief, perhaps, was that timber could generally be obtained in sufficient quantities, although in some of the older regions with the passage of time it became somewhat scarce. The prejudice against iron made with anthracite coal or with coke, the difficulty in discovering coal suitable for coking and for smelting iron satisfactorily, the lack of good transportation facilities before railroad development, and the inertia due to custom constituted other important reasons.

A strong argument in favor of the continued use of charcoal was that it made an ideal furnace fuel. It was almost free from sulphur, and its ash consisting largely of lime and alkalis, supplied a part of the flux required in the process of smelting the ore. The ash of coke being largely silica, has itself to be fluxed with different bases. On the other hand, when coke and charcoal are compared in blast furnace use, there are differences in favor of coke. Charcoal does not have the physical strength of coke and cannot be used in the exceedingly high furnaces of the present day. The chief reason for the change from charcoal to coke, however, was the unsurmountable difficulty of wood supply. Even the vast forests of America could not have continued to supply the demand as the iron industry grew and developed. It should be noted that wood in its raw state, containing a large amount of water and soluble minerals, was never considered a suitable material for smelting iron. Since the change in fuel from charcoal to coke in Pennsylvania iron smelting did not occur until after the period under consideration in this volume the question will be left for discussion in a succeeding work.

The process of charcoal making was an interesting one and the picturesque aspects of this phase of human activity have not been neglected by writers of literature. It required many colliers to "coal" the wood. Drinker estimated that as many as twelve colliers might be necessary to keep a single furnace going.[33] Usually two or three master colliers worked with the aid of several helpers. When the piles were in process of charring, it was necessary to watch them carefully day and night. Thus the bleak and lonely charcoal burners' huts were

built in the silent forests or woods, far away from the plantation community where the rest of the ironmaster's workers
lived.

A dry, level spot near the woodland, sheltered from the
wind by a declivity or by trees, was selected. The ground had
to be dry and fine, free from stones and gravel, but not loamy
or sandy. A circular spot from thirty to fifty feet in diameter, called the "pit" or "hearth," was prepared. Prior to this

Charcoal Pile in Process of Being Charred

the trees had been felled, trimmed, cut into lengths, and
bundled into cords, by the woodchoppers. This wood was
brought to the "pit," and the colliers stacked it in conical
shape, standing the sticks on end. This was done by first forming in the center a circle of sticks to compose the chimney,
three or four inches in diameter and about six feet high.
Around this chimney the cords of wood were arranged to form
the cone which was about twenty-five feet in diameter at the
base.

After the cone was built, a layer of damp leaves and loose
earth, or a coating of turf was placed on the pile. Chips, dry
leaves or other inflammable material were placed into the
chimney and lighted from the top, which was then partly
closed with turf. Holes were drilled in the side to draw air.
After the pile was lit, it had to be carefully watched day and
night, and any sign of flame smothered. It took from three to

ten days to char the entire pile, depending on the nature of the wood, the state of the weather and the skill of the colliers. In European practice, the form of the pile was sometimes pyramidal or rectangular,[34] but in Pennsylvania and generally in America the form was that of circular cones.[35]

After the pile was completely charred, the colliers raked the charcoal into small piles, so that if any should break out into a blaze through spontaneous combustion, the whole mass would not be ignited. Hot, dry charcoal absorbs oxygen so rapidly that this danger is always present. The product was then hauled to the thick-walled stone charcoal house always found near the furnace or forge where the charcoal was placed to be ready for use.

Various woods were used in charcoal making including hickory, white oak, black oak, ash, chestnut and pine. Hickory was considered the best, but as black oak was abundant, it was more generally used.[36] The yield of charcoal varied considerably, depending upon the quality of the wood, the more compact and fine-grained giving the largest yield.

The amount of charcoal consumed by the furnaces was enormous. Oley Furnace in 1783 consumed about 840 bushels every twenty-four hours, for which about twenty-one to twenty-two cords of wood were necessary. This furnace produced a little more than two tons of iron a day and therefore for every ton of iron more than 400 bushels of charcoal were required.[37] Since an acre of land is necessary to produce about twenty-five cords of wood of twenty-year to twenty-five year growth,[38] Oley Furnace used every day almost the amount of wood grown on an acre. In 1786, Drinker estimated that an average furnace of his time would consume 800 bushels of charcoal every twenty-four hours.[39] The famous Warwick Furnace used from 5,000 to 6,000 cords of wood during the period it was in blast each year. This was the product of about 240 acres or more of woodland.[40]

Many men were required to cut down the forests to provide the charcoal burners or colliers with wood. This work was usually done during the winter months. Most of the furnaces were shut down for at least two or three months each year for repairs or relining. Difficulties resulting from the freezing of the water supply also often forced a shut-down during severe weather. Most of the workers, including the furnacemen and miners generally worked with the woodchoppers. After the wood was cut, it was not charred immediately. In fact the charcoal was not made until a short time before it was needed. The strongwalled charcoal houses, large as most of them were, could not contain the amount of charcoal necessary to feed the furnace for a great length of time, and to have left it outside would have made it unfit for use in the furnace.

The various processes of iron manufacture in eighteenth century America were comparatively simple when compared with the highly complex processes of the iron and steel industry today. The first ironworks in Pennsylvania as in many other colonies were bloomery forges. They were patterned after bloomeries scattered over Europe during this period. These had their origin in the Catalan forges which had developed in the hills of Catalonia, Spain, about the tenth century.[41] During the centuries that followed, the manufacture of iron developed rapidly in many parts of Europe, for new uses were found for iron. The extended use of iron was aided by the Catalan process which was a marked improvement over the earlier primitive methods of working iron in holes on hills of clay, or in small hearths, where the wind or a crude goatskin bellows worked by hand or foot furnished the blast. In many ways the early bloomeries in Pennsylvania resembled the Catalan forge. They were several times larger than the ordinary blacksmith forge of the present day, and had deep, wide fire-pots. A blast obtained by means of water-power bellows was introduced at the side. The fuel was charcoal.

The ore from nearby mines was broken into small pieces and put into the heated hearth of the forge. As the iron became semi-molten, the forgeman stirred or worked it with a long bar until it gathered into a lump, called a *bloom* or *loupe*. The blazing ball of crude iron, pasty but never molten, was taken out and placed under the ponderous hammer operated by a water wheel.[42] Under the rhythmic strokes of the hammer, a heavy bar was fashioned. Successive reheatings and hammerings produced bars for the use of the blacksmith or artisan. This iron was a type of wrought iron. The process of making iron in bloomeries was very wasteful. Large quantities of charcoal were used. The iron that was made contained much slag and therefore was very impure. There were only a few works of this type in Pennsylvania, the blast furnace instead coming into use from almost the very beginning of the industry here. The reverse was true in New England, where bloomeries flourished, especially, throughout the latter part of the colonial period.[43]

The blast furnace of the eighteenth century was not a sudden invention, but the result of a slow evolution. The Catalan forge or bloomery had been improved and enlarged during the Middle Ages. Then followed the German stuckofen which was an attempt to utilize the waste heat of the bloomery and was built about twelve feet or more in height. In this, pasty malleable metal or freely flowing molten cast iron could be produced. The stuckofen did not become popular, nor did it succeed in displacing the bloomery.[44] The next step was the "high furnace" or blast furnace. Just when this transition took place is uncertain. Scattered bits of evidence point to the fourteenth century.[45] It is certain that in the days of Henry VIII there were many blast furnaces in England.[46] By the eighteenth century, they had become somewhat standardized and both in England and on the continent they were built about twenty-five to thirty feet high. The English ironmasters who built the first furnaces in Pennsylvania modeled them after those of the mother country.[47]

The old furnaces were built into the side of a small hill in order that the ore, limestone flux and charcoal could be put into the stack from the top. Otherwise some means would have been necessary to hoist the materials to the furnace top. The stacks were built square, but they tapered, being larger at the bottom than at the top. They varied in size, but were usually about twenty-five feet square at the bottom, and from twenty-five to thirty feet high.[48] The earliest furnaces were entirely open at the top; later ones had a cylindrical erection of brickwork over the tunnel-head or throat of the furnace, for the protection of the workers from the heated gases and smoke rising from the furnace. Through a door in the tunnel-head the charges of material were made. The tunnel-head was connected with the bank by a wooden bridge over which the "fillers" continually crossed bearing their baskets loaded with material to feed the hungry furnace.

Down below, in front of the furnace, protected by a roof or building of massive timbers, was the casting house or casting shed, where the molten metal from the hearth ran into the sand molds. The entrance to the hearth was through an arch in the furnace wall. Outside the shed, on the one side of the furnace, was an arched recess in which a small aperture allowed for the insertion of the "tue-iron" (tuyère), and also the iron pipe connected with the large water-driven bellows which furnished the blast. The eighteenth century furnaces

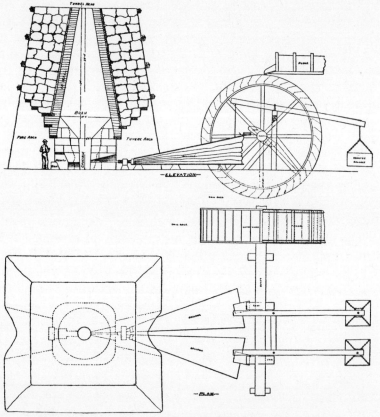

Plan of Eighteenth Century Cold-Blast Charcoal Furnace

had only one tuyère, which was open and led directly into the hearth. There were no water coils or any other cooling appliances as can be found in modern furnaces.

The outside portion of the stack was usually built of large blocks of limestone or other type of stone found in the locality. The interior was lined with a reddish fine-grained sandstone, such as was obtained in the Schuylkill Valley for the furnaces of southeastern Pennsylvania. A similar stone was also obtained at Lebanon.[49] Wacke[50] and slate were sometimes used, after the middle of the century, but not always with great success.[51] Between the inwalls and the outside limestone, a space of a few inches was filled with clay or coarse mortar to protect the outside limestone from the decomposing effects of the heat. Some of the stacks were re-enforced and strengthened with strong iron girders embedded in the walls. The widest part of the inner chamber of the stack—the bosh—was usually nine feet in diameter. The hearth, a cylindrical reservoir at the bottom of the furnace, into which the molten metal ran, was relatively small, being only a few feet in diameter, owing to the necessity of concentrating the molten iron to prevent it from solidifying.

It is quite evident why the bosh—the central part of the stack—was made wider than the upper and lower interior parts of the furnace. When, in the evolution of the furnace, its height was increased the pressure of the superincumbent mass within, rendered the materials so dense as to retard the ascent of the blast. Hence the internal buttresses or boshes were introduced to support the weight of the charge, relieving the central part from pressure and permitting the free passage of the blast. This principle was used in the stuckofen. While the good quality of the iron and the regularity of the process was thus assured, further increased output came by the use of larger and thus more powerful blowing apparatus. The output of the furnace during the latter part of the century averaged about twenty-five tons weekly,[52] although Warwick, an exceptional furnace, often made forty tons a week before the close of the Revolution.[53] The first furnaces established in the twenties made from fifteen to twenty tons weekly.

The operation of the furnace—a chemical operation—was fairly simple although its management often involved great difficulties. It was kept filled continuously with alternate layers of charcoal, ore and limestone.[54] These materials were usually not weighed before being charged but simply measured in baskets or buckets. The practice of weighing began with the use of anthracite coal in the smelting of iron which came much later than the period under discussion. The gaseous products escaped at the top of the stack. At the tuyère of the furnace the ore melted and dropped down to the hearth below. The cinder or slag floated on top of the molten iron and was drawn off from time to time.[55] The slag was the result of the action of the limestone flux upon the impurities of the ore. About twice a day, sometimes oftener, the molten iron was run into the casting bed of sand,[56] which was prepared for its reception by molds occasionally made from mahogany wood patterns,[57] but usually from other wood patterns. Some imaginative early ironmaster compared the casting bed to a sow and her litter of sucking pigs. Thus the main stream or feeder from the furnace was called the sow, while the side gutters were called pigs, a term which is in use today. Before the iron became cold, the pigs were separated from the sow and the latter broken up into smaller pieces. The amount of ore necessary to make a ton of pig iron varied, depending on the type of ore, the construction of the furnace and skill of the manager. About two tons of ore usually made a ton of cast iron.

The "blowing in" or the starting of a furnace was difficult. The stack was first filled with charcoal and lighted from the top. After several days, when the fire had burned down and reached the tuyère opening, the furnace was re-filled with charcoal. The fire now worked back to the top. The blast was then applied, and ore and flux put in from the tunnel head in gradually increasing quantities. After a few days slag and iron ran into the hearth below. The proportion of ore and flux to charcoal was gradually increased until the furnace was working normally. The slag varied in color. That of a fine sky-blue color denoted the presence of manganese. Gray slag indicated high grade iron, rich in graphite carbon. Dark slag showed that the iron was low in graphitic carbon.

Besides pig iron, the early Pennsylvania furnaces cast hollow ware such as pots, pans, skillets, sugar kettles, Dutch ovens, stoves, and firebacks. The process was similar to that of casting pig iron. Different molds, of course, had to be used. When the molten iron was tapped into the pig beds, a part of it was delivered into large ladles, which in turn was poured into small ladles and then into the molds for castings. Many of the decorated stove plates with their biblical and classical scenes, scriptural quotations, mottoes, hearts and flowers,[58] still remain, bearing testimony to artistic yearnings of these early artisans in iron. The verses and mottoes in a form of

German and the Germanized spelling of English names reveal the origin of many of these early workers.

The burdening of the furnace—the determination of the proper proportions of ore, flux and fuel for the furnace charge —was considered a "gift" of the old time furnace worker. The results obtained in view of lack of knowledge, were remarkable. The modern laboratory method with its quick and accurate analyses has done away with such guess work, or trial and error, but there is still the necessity for judgment, experi-

Stove Plate

ence and a certain intuition which successful furnacemen must have.

The number of workmen needed to operate the furnace was not large. Two founders, two keepers, two guttermen, two or three fillers, who filled the furnace with alternate charges of charcoal, ore and limestone, a "potter," who made the hollow ware, an ore roaster and a few laborers included them all.⁵⁹ The "founders," who regulated the furnace, made the sand molds, and cast the iron, together with the "potter" were the only skilled workmen employed at the furnace. Frequently the "potter" was an itinerary worker, working a few weeks at one furnace and then traveling on to another. As the work of ironmaking had to be carried on night and day, the workers labored in two twelve hour shifts.

While the length of time a furnace was in blast did not

usually exceed nine months,[60] some furnaces like Reading Furnace occasionally ran from twelve to eighteen months at a stretch.[61] The furnaces were generally shut down during the winter months,[62] and work was suspended at times during the heat of the summer.[63] While the furnaces were out of blast, there was much to be done. The furnace had to be relined, occasionally rebuilt; roads, dams, and bridges repaired; swamps filled in; and above all, wood had to be cut in preparation for the next blast.[64]

The blast required for the furnaces and forges was obtained by means of large double bellows made of wood and leather, always driven by huge water wheels. The bellows of even the earliest Pennsylvania furnace, were large, measuring from twenty to twenty-five feet in length and several feet in width. Hermelin gives a clear picture of how the blast was secured:

> At most of these blast furnaces there is no other water supply than from small brooks and springs, which is conveyed to a dam between the mountain heights, where the water goes to the blast furnace wheel, partly by [means of] a ditch and partly through a grove made of hollowed out tree trunks and also through wooden pipes. At most blast furnaces there are overshots for the wheels, which have a diameter of forty-four feet and can therefore be driven by a small amount of water.[65]

While bellows were chiefly used during the eighteenth century at the blast furnaces, long before the century closed blowing cylinders or blowing tubs were installed in many of the ironworks, especially at the forges.[66] These were the invention of James Knight of Bringewood, England, who obtained a patent for them in 1762,[67] and John Smeaton, who used his blowing cylinders at the famous Carron Iron Works.[68] This new method of providing blast soon came into favor with other English ironmasters, and displaced the simple wood and leather bellows in England. Of the many inventions in the technique of iron making at blast furnace and forge brought about in England during the last half of the eighteenth century, the blowing tubs alone were used in America, and not a very large number of ironmasters in this country substituted these for the bellows prior to the close of the century.

In Pennsylvania the blowing tubs, including cylinders, piston and connecting rods were made of wood. The earliest ones were crude, being little more than two cylindrical casks fitting closely into one another, moving up and down between four wooden posts, the motive power being a large water wheel. The air, blown into a leather bag, was conveyed to the furnace in an iron pipe.[69] After many improvements, the so-called double cylinders came into use. These consisted of two wooden

Hay Creek Forge, Front View

cylinders set side by side. These blew into a third cylinder which was ordinarily weighted, so as to get as uniform a pressure as possible,[70] thus overcoming the insufficient and unsteady blast of the old bellows.

From the time cast iron was first made in the fourteenth century or earlier, some process had to be devised for refining it. Thus refinery forges came into existence. Like the bloomeries and blast furnaces, the refinery forges of Pennsylvania were patterned after those of England, and the methods employed were English. At a few Pennsylvania forges the Walloon method, which was quite similar to the English procedure, was used. The forges varied in size. Small ones had but one "fire" or "stack" (hearth) and one hammer; the largest boasted of four "fires" and two hammers.[71] Pig iron in rough bars from five to six feet long and about six inches broad was used.[72] Two or three were put into the first hearth, called the "refinery," one end being placed in the charcoal fire. As the ends softened under the heat of the blast in the deep fire-pot, the portions of the bars outside the fire were pushed in. The "finers" worked the mass with long iron bars into a lump called a "half-bloom." This process was similar to that of making "blooms" at the bloomeries.[73] The mass was slung by means of hooks and tongs from the hearth and placed under a weighty hammer, driven by a water wheel in order to refine it or to hammer out its impurities. The first process in the fire had done much to drive out the carbon. The half-bloom, placed in the "refinery" once more and heated to a bright-red color, was worked under the hammer again, this time into an "ancony"—a flat, thick bar, with a rough knob on each end. These were sometimes sold in the market as bars of commerce. But more often, the "ancony" was re-heated at another hearth, called the "chafery" and worked under the hammer into long bars at the same forge.[74] The bars were then cut into convenient length, and were ready for sale to blacksmiths, locksmiths and others to be made into finished products of various kinds. The finished bars varied in length, width and thickness. Often they were drawn about fourteen feet long, two inches broad and one-half an inch thick; others were drawn several feet long and one inch or two inches square.[75]

The forge hammer or hammers were massive. Each hammer-head was of a heavy piece of wrought or cast iron weighing several hundred pounds. It was mounted on an oak beam and usually ran parallel to the water-wheel shaft, the trunnions of which caught the helve or oak beam a little distance behind the hammer head.[76] The foundation of timber and stones bore the weight of the heavy strokes on the resounding anvil. Water wheels measuring twenty-five or more feet in diameter furnished the power for the hammers. Even larger water wheels provided the power for the wooden and leather bellows or for the "blowing tubs" used at many forges during the latter part of the century, which furnished the blast for the "fires."[77]

Skilled men worked at the hammers and at the hearths. The iron was first worked by the "finers." The technique of swinging the half-bloom to the hammer and back to the hearth required much strength and practice. The hammermen, likewise, were experienced. It required no little degree of strength to draw the bars to exact given sizes. In addition to the three or four skilled workers, a few laborers and perhaps one or two apprentices completed the number who worked at each forge.[78]

The output of bar iron at the forges varied. The smaller ones produced about two tons a week. A forge with three or four hearths and two hammers could turn out much more, and such a forge working double-handed often made from 300 to 350 tons annually.[79] Charming Forge, for instance, even prior to the Revolution had a capacity of 300 tons a year.[80]

The bar iron produced at forges was shaped into finished products chiefly by blacksmiths, who occupied a very important position in iron manufacture during this period. There were hundreds of blacksmith shops, large and small, in Pennsylvania, as in other regions of America where bars of iron were wrought into tire iron, axes, hoes, shovels, chains, scythes, and other needed articles. The blacksmith at this time was a skilled artisan and a craftsman; his activities were not confined to repair work.

At many of the ironworks, stamping mills were located, especially during the latter part of the century. These crushed forge cinders and furnace slag in order to obtain the iron which they contained.[81] Much iron was recovered in this way and was used again at the blast furnaces or bloomeries. In spite of all efforts, the loss of iron in the scoria or slag was high, especially when the furnaces were not working as well as they should.

The amount of steel made in Pennsylvania during the eighteenth century was comparatively small. Most of it was "blister" or "cemented" steel, although toward the end of the century, the German method of making steel direct from pig iron was attempted.[82] The demand for steel was not great because it was used only for swords, bayonets and edged tools, and not for the manifold uses to which it is put today. The high cost of steel prevented its extended use.

The capacity of steel furnaces varied from three to ten tons,[83] although the new furnace in Philadelphia, which George Washington visited in 1787[84]—the largest and best constructed in America—had a capacity of fourteen tons. These furnaces, built about twelve feet high had a fire grate above which were two long pots built of stone or brick. Into the pots which were about four feet wide, three feet high and about sixteen feet long, were placed forged iron bars with alternate layers of small sized charcoal or charcoal dust.[85] After the pots were filled to the top, the fire was lighted. The furnace was kept at a cherry heat from seven to eleven days, depending upon the hardness of steel desired. The carbon from the charcoal dust was absorbed by the bars and they were transformed into steel. There was a small opening in each pot for a proof bar, which was taken out and tested from time to time. When the operations were thought to be complete, the

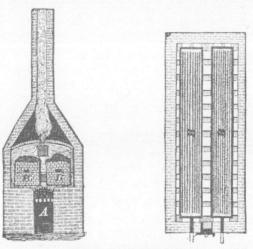

Blister Steel Furnace

fire was put out and the furnace left cool. The steel bars were then ready for the market. Crucible steel, made first in England by Benjamin Huntsman, a clockmaker who was determined to make springs of a better grade for his clocks, in 1740,[86] was not successfully made in America until the nineteenth century. Only small quantities were made in this country before the Civil War.

By the end of the eighteenth century, air furnaces had been introduced into Pennsylvania, especially in the region of Philadelphia.[87] These re-smelting furnaces, having a capacity up to approximately five tons for making castings of all kinds from pig iron, were the progenitors of the modern cupolas.

They produced loom, dry sand, flask, and open sand castings consisting of forge hammers, slitting mill rolls, stoves, anvils, skillets, pots, pans, kettles, and other types of castings.[88]

During the period under discussion sheet iron and tin plate

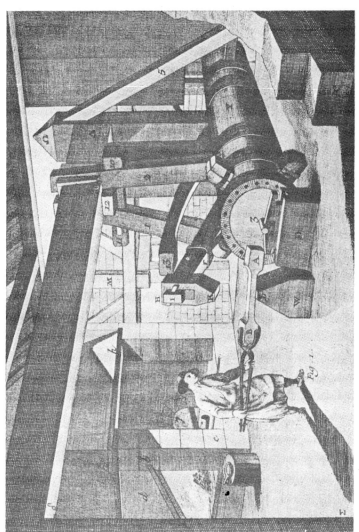

Forging an Ancony at Refinery Forge

iron were never rolled, but were hammered out in Pennsylvania at the tilt-hammer or plating mill. Such a mill had a large hammer that worked much faster than the helve hammer of the forge. The smaller size of the water wheel and the large number of cogs in the wheel caused the hammer to give very rapid strokes. Bar iron, brought from the forge, was heated and hammered into sheets. Iron sheets had been rolled in England as early as 1728, an invention claimed by Major Hanbury and John Payne. These early experiments of rolling plates evidently were satisfactory, the plates being pliable and of a uniform gage. It is remarkable that this method was not adopted generally in England, but no further progress was made until after Henry Cort invented grooved rolls in 1783-84.[89] No plates, however, were rolled in the United States, until many years after the opening of the nineteenth century.

At the rolling and slitting mills, slit iron was prepared to be used for making nails. The iron bars received from the refinery forge were cut into strips by a water-power crocodile shears. The strips, heated to a red heat, were passed through "rollers" (rolls) until they became the thickness of the intended rod. After being reheated, each strip was presented to the slitters, which consisted of small grooved rolls, set so that the rims of one roll entered the grooves of the other. In the "slitters" the strip was divided into several rods called slit iron. Large water-wheels drove the bellows in the hearths, the rolls, and shears. Slitting mills and nail works were often found in combination, but much slit iron was sold to ironmongers, blacksmiths and farmers who manufactured nails themselves.[90]

During the latter part of the century, many nail works sprang up in Philadelphia and in the various boroughs and towns where machines for cutting and heading nails were in use.[91] These works produced nails of all sizes. They were equipped with anvils, small forges, vises and nail-making machines of various kinds. Frequently, as many as twenty-four or more hands were employed in a single plant,[92] foreshadowing the factory system that was soon to appear.

Iron wire was one of the few articles that had to be almost entirely imported during the eighteenth century. Several unsuccessful attempts were made to establish wire mills. A wire drawing plant was set up in Philadelphia as early as 1775.[93] At these works thin strips of iron were attached to a machine worked by water power, which drew them through a "drawplate." Unlike the process of forging and rolling, the drawing of wire, even today, is still done with cold metal. It was not until the early years of the nineteenth century that a wire

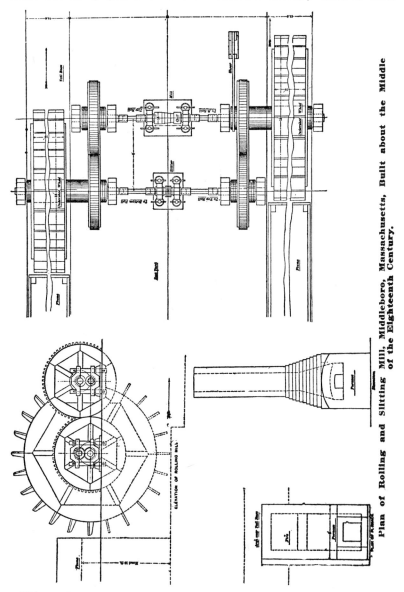

Plan of Rolling and Slitting Mill, Middleboro, Massachusetts, Built about the Middle of the Eighteenth Century.

mill run by steam power was established in Pittsburgh, by means of financial aid from the state legislature. This marked the real beginning of the wire industry in the state.[94]

Tin plate works where tin plate was worked into vessels of various kinds, flourished during the latter half of the century. White smiths, the workers in tin plate, established shops in towns and boroughs. Twenty or more journeymen were employed in many of these shops.[95] A variety of utensils such as kettles, coffee pots, saucepans, fish kettles, Dutch ovens, and stew pans, were fashioned from plate iron and sheets produced at the tilt hammers or plating forges and from imported tin plate.[96] The name whitesmith was applied to workers of tin plate.

Various types of iron were manufactured in the eighteenth century. The iron of antiquity—wrought iron—was the pioneer of the iron family. When early civilization began to make progress wrought iron was made by early man on the tops of hills where brisk breezes provided the blast, the forest trees supplied the fuel, and large stones were used as hammers. After the passage of centuries wrought iron was made in low hearths, bloomeries and forges, but was always hammered out. In addition to the iron itself this type of wrought iron contained some viscous cinder or slag, but little or no carbon. In the hammering out of the bloom or bar, the slag was extended lengthwise throughout the bar. Thus wrought iron even in its crudest form, because of its fibrous structure possessed malleability, ductility, tenacity and toughness. A microscopic examination of such iron reveals filaments of slag lengthwise in the bar. The hammering did not alter the chemical composition, except to eliminate a certain amount of impurities, but it changed the physical structure by reducing the impurities to lengthy filaments or fibers, thus strengthening the texture of the iron.

With the introduction of the principle of the "high furnace" or blast furnace toward the close of the Middle Ages, having as its object the utilization of the waste heat of the bloomery forge, and an increased output, a different type of iron is found. In its journey through the blast furnace, the iron being molten absorbed from three to five per cent of carbon. This carbon was in two forms: (1) the combined form, that is, in chemical combination with the iron itself, and (2) as graphite flakes amid the particles of iron. The same is true of all cast iron made today. A microscopic examination shows black flakes of free carbon distributed throughout the alloy. This is the reason for the brittleness of cast iron.

Iron which contains from 0.25 to 1.8 per cent of carbon in chemical or molecular combination (not graphite or free carbon) is called steel, and was the third type of iron known and made in the eighteenth century. The famous steels of the past —the "Wootz" of India, the "Damascus" of Syria, and the "Toledo" of Spain were made from wrought iron by impregnating them with the correct percentage of carbon generally by the trial and error method.

The conversion of bar iron into steel by placing it in molten metal for several hours is generally ascribed to Reaumur, but this process is mentioned by Agricola, writing in the middle of the sixteenth century,[97] more than one hundred and fifty years before Reaumur. The latter, however, was the first to understand to any extent the cementation process of placing bars of iron together with powdered charcoal into ovens which were kept at cherry heat for many days, the bars absorbing the carbon from the charcoal and becoming steel. The publication of his complete directions for making steel by this process, about 1722,[98] gave a great impetus to the manufacture of steel. France, his own country, however, was unable to profit very much by his discoveries because of her small production of suitable iron. Sweden, England and Germany were greatly benefited, and cementation furnaces were also built in America. Several were built in Pennsylvania during the eighteenth century.

Cementation steel is called "blister steel" because it shows blisters on the surface and contains blisters throughout. The reason for these blisters was not discovered until 1864 when John Percy, the English metallurgist, suggested that they were caused by the chemical action of the carbon on the slag contained in the wrought iron.[99] The gases formed, he said, produced the blistering of the bars. This explanation is proved by the fact that bars of mild steel, or iron without slag, such as are made today, do not blister when subjected to this process. Although the Huntsman, Bessemer, Open Hearth, and other processes for making steels of various kinds have supplanted the old methods, the best steel used in the manufacture of cutlery is still made by the cementation method especially in Sheffield, England.

Pennsylvania was the foremost iron producing center

throughout the latter half of the eighteenth century.[100] Today the state is one of the greatest iron and steel producing districts in the world. The location of the industry has changed. In the eighteenth century it centered chiefly in southeastern Pennsylvania among the rich ores of the state, although there were industrial regions even to the western borders by its close. It is now concentrated in the Pittsburgh district. This has been due to several reasons. During the westward movement, Pittsburgh was the key to the West, and after the industry crossed the Alleghenies and established itself there, the ironworks of the Pittsburgh district furnished most of the iron for the agricultural implements and tools needed in the growing West. When coke displaced charcoal in blast furnace production, an impetus was given to the industry in this district because the soft coals of the Connellsville region were found to be the best for making coke. The discovery and the relatively cheap transportation of the Lake Superior ores; the excellent sites for mills along the rivers; the progressiveness of the ironmasters and manufacturers in using and developing the Bessemer, Open Hearth, and other processes and inventions, and in overcoming innumerable obstacles all combined to build up the great industrial regions of the present day.

FOOTNOTES

[1] F. W. Clarke, *The Data of Geochemistry,* United States Geological Survey, *Bulletin* 491, p. 33.

[2] Classification of the United States Geological Survey.

[3] Samuel G. Gordon, *Mineralogy of Pennsylvania,* p. 52.

[4] A. C. Spencer, *Magnetite Deposits of the Cornwall Type in Pennsylvania,* United States Geological Survey, *Bulletin* 359, pp. 107 ff.

[5] Israel Acrelius, *History of New Sweden,* p. 169; Henry Horne, *Essays Concerning Iron and Steel,* p. 23.

[6] Several attempts were made in England as well as in America to produce iron in blast furnaces from the magnetic sands found along the New England and Virginia tidal rivers. See Henry Horne, *Essays Concerning Iron and Steel,* pp. 23 ff.

[7] A. C. Spencer, *Magnetite Deposits in Berks and Lebanon Counties.* United States Geological Survey, *Bulletin* 315, pp. 185-189.

[8] Samuel G. Gordon, *Mineralogy of Pennsylvania,* p. 50.

[9] E. C. Eckel, *Iron Ores,* p. 259; J. J. Rutledge, *Clinton Iron Ores,* American Institute of Mining Engineers, *Bulletin XXIV,* 1908.

[10] Samuel G. Gordon, *Mineralogy of Pennsylvania,* pp. 55-60.

[11] E. C. Eckel, *Iron Ores,* pp. 259-260; T. C. Hopkins, *Cambro-Silurian Limonite Ores of Pennsylvania.* Geological Society of America, *Bulletin* 1900, II, 475-502.

[12] Samuel G. Gordon, *Mineralogy of Pennsylvania,* p. 64.

[13] E. F. Burchard and H. W. Davis, *Iron Ores, Pig Iron and Steel Mineral Resources of the United States,* 1924. Department of Commerce Publications. Part I, p. 299.

[14] Israel Acrelius, *History of New Sweden,* p. 169.

[15] C. D. Ebeling, *Die Vereinten Staaten von Nordamerika,* IV, 89.

[16] Peter Kalm, *Travels into North America,* I, p. 236.

[17] J. D. Schoepf, *Travels in the Confederation,* II, 7.

FOOTNOTES—Continued

18 Israel Acrelius, *History of New Sweden*, p. 165; J. D. Schoepf, *Travels in the Confederation*, I, p. 202.
19 Israel Acrelius, *History of New Sweden*, pp. 164-165.
20 J. D. Schoepf, *Travels in the Confederation*, I, 207-208; S. G. Hermelin, *Report About the Mines in the United States of America, 1783*, p. 15.
21 J. D. Schoepf, *Travels in the Confederation*, I, 197-198; S. G. Hermelin, *Report About the Mines in the United States of America, 1783*, p. 28.
22 J. D. Schoepf, *Travels in the Confederation*, I, 198.
23 *New Jersey Gazette*, May 12, 1781.
24 J. D. Schoepf, *Travels in the Confederation*, I, 202.
25 Henry Drinker to Richard Blackledge, 10th month 4th, 1786, Drinker Letter Book, 1786-1790, p. 83.
26 35 Henry VIII; c. 17; 1 Elizabeth, c. 15; 23 Elizabeth, c. 5; 27 Elizabeth, c. 19.
27 35 Henry VIII, c. 17; T. S. Ashton, *Iron and Steel in the Industrial Revolution*, p. 15.
28 *Naturalis Historiae*, Book XVI, c. 6.
29 D. Mushet, *Papers on Iron and Steel*, p. 69.
30 Dud Dudley, *Mettallum Martis*.
31 Compare T. S. Ashton, *Iron and Steel in the Industrial Revolution*, pp. 29-36, Appendix E with H. Scrivenor, *History of the Iron Trade*, p. 55.
32 T. S. Ashton, *Iron and Steel in the Industrial Revolution*, p. 37.
33 Henry Drinker to Richard Blackledge, 10th month 4th, 1786, Drinker Letter Book, 1786-1790, p. 83.
34 *Encyclopedia Britannica*, 1797. Article on "Charcoal."
35 Even today farmers frequently dig up the remains of the circular pits. In the pine regions of southern New Jersey, charcoal is still made by the old method.
36 Israel Acrelius, *History of New Sweden*, p. 168; J. D. Schoepf, *Travels in the Confederation*, I, 198; *Alden's Appeal Record*, Coleman *vs.* Brooke, p. 309. A traveler in the twenties stated that black walnut, black oak and ash were the woods chiefly used at that time. Quoted in S. W. Pennypacker, *Hendrick Pannebecker*, p. 84.
37 J. D. Schoepf, *Travels in the Confederation*, I, 198.
38 *Alden's Appeal Record*, Coleman *vs.* Brooke, p. 145.
39 Henry Drinker to Richard Blackledge, 10th month, 4th, 1786, Drinker Letter Book, 1786-1790, p. 83.
40 Mrs. T. P. James, *Memorial of Thomas Potts, Jr.* p. 47.
41 M. J. Francois, *Researches*, pp. 404 ff., Quoted in John Percy, *Metallurgy*, pp. 278 ff.
42 Emanuel Swedenborg, *Regnum Subterraneum sive Minerale de Ferro*, Part I, Section 13.
43 William Douglass, *British Settlements in North America*, II, 109.
44 M. Jars, *Voyages Metallurgiques*, I, 37 ff; John Percy, *Metallurgy*, pp. 326 ff.
45 John Percy, *Metallurgy*, pp. 878-879; Georgius Agricola, *De Re Metallica*, p. 420.
46 T. S. Ashton, *Iron and Steel in the Industrial Revolution*, p. 4.
47 Emanuel Swedenborg in his *Regnum Subterraneum sive Minerale de Ferro*: Part I, Section 13, states that the height of the early Pennsylvania furnaces, which he inspected was twenty-five feet.
48 See *Appendix* B for the height of eighteenth century Pennsylvania furnaces.
49 J. D. Schoepf, *Travels in the Confederation*, I, 198, 208.
50 Rock similar to sandstone in texture, but derived from disintegrated basic rocks.
51 J. D. Schoepf, *Travels in the Confederation*, I, 198.
52 See *Appendix*, C and D.
53 J. D. Schoepf, *Travels in the Confederation*, I, 202. Warwick and Reading Furnaces produced 800 tons of pig iron annually even before the middle of the century, which was equal to the output of the largest English furnaces of the same period. Israel Acrelius, *History of New Sweden*, p. 168.
54 From two to three hundredweight of limestone was used each day by the average furnace to flux the ores of their impurities. Henry Drinker to Richard Blackledge, 10th month 4th, 1786, Drinker Letter Book, 1786-1790, p. 83.
55 *Alden's Appeal Record*, Coleman *vs.* Brooke, pp. 205, 221. The journey of the ore through the furnace from the time it was put in until it reached the hearth, took from forty to sixty hours.
56 *Ibid.*, pp. 160, 165.
57 J. D. Schoepf, *Travels in the Confederation*, I, 199.
58 The collection of stove plates at Doylestown is described in H. C. Mercer, *Decorated Stove Plates of the Pennsylvania Germans* and in his *Bible in Iron*.
59 Henry Drinker to Richard Blackledge, 10th month 4th, 1786, Drinker Letter Book, 1786-1790, p. 83.
60 See *Appendix*, C and D.
61 J. D. Schoepf, *Travels in the Confederation*, I, p. 202.
62 *Alden's Appeal Record*, Coleman *vs.* Brooke, p. 203.
63 Israel Acrelius, *History of New Sweden*, p. 168. Acrelius states that all work at the furnaces and forges was suspended during the hot summer months.

FOOTNOTES—Concluded

This was not always true, as is evidenced by the Furnace and Forge Ledgers. See also Mrs. T. P. James, *Memorial of Thomas Potts, Jr.*, pp. 44 ff, 48-49.

64 *Alden's Appeal Record*, Coleman *vs.* Brooke, p. 203.

65 S. G. Hermelin, *Report about the Mines in the United States of America, 1783*, p. 59. Compare with Emanuel Swedenborg, *Regnum Subterraneum sive Minerale de Ferro*, Part I, Section 13 ; James Thacher, "Observations Upon the Ores." Massachusetts Historical Society, *Collections*, First Series, IX, 259.

66 S. G. Hermelin, *Report About the Mines in the United States of America, 1783*, p. 63 ; *Alden's Appeal Record*, Coleman *vs.* Brooke, p. 205.

67 T. S. Ashton, *Iron and Steel in the Industrial Revolution*, p. 37.

68 H. Scrivenor, *History of the Iron Trade*, p. 82.

69 J. D. Schoepf, *Travels in the Confederation*, II, 6 ; *Cazenove Journal*, pp. 6 ff.

70 *Alden's Appeal Record*, Coleman *vs.* Brooke, p. 205.

71 Israel Acrelius, *History of New Sweden*, pp. 165 ff. *Philadelphia Weekly Mercury*, March 10, 1768, March 9, 1769, November 23, 1791, etc.

72 Israel Acrelius, *History of New Sweden*, p. 168.

73 Emanuel Swedenborg, *Regnum Subterraneum sive Minerale de Ferro*, Part I, Section 13.

74 Israel Acrelius, *History of New Sweden*, p. 168 ; Henry Horne, *Essays on Iron and Steel*, p. 90 ff.

75 Eighty-six to ninety bars usually weighed a ton. Potts MSS. XL, Pottsgrove, 1755 (1755-1758), pp. 185-186.

76 J. D. Schoepf, *Travels in the Confederation*, II, 6. It appears that most, if not all the forge hammers of this period were similar to the one at Coventry, described by Schoepf. The recent excavations at Valley Forge revealed the remains of a hammer of the same type.

77 S. G. Hermelin, *Report About the Mines in the United States of America, 1783*, p. 63 ; James Thacher, "Observations on Iron Ores," in Massachusetts Historical Society, *Collections*, First Series, IX, 259.

78 Israel Acrelius, *History of New Sweden*, p. 168 ; Potts MSS. XL, Pottsgrove, 1755 (1755-1758) p. 193 ; *Ibid.* LXX, Pine Forge, 1774 (1774-1781) p. 59.

79 Board of Trade Papers: Proprieties, X, W23 ; Israel Acrelius, *History of New Sweden*, p. 168 ; Report to Governor Franklin, *New Jersey Archives*, First Series, XXVIII, 247.

80 *Pennsylvania Gazette*, May 10, 1770.

81 *Reading Weekly Advertiser*, November 19, 1796.

82 *Philadelphia Federal Gazette*, February 13, 1799.

83 Henry Horne, *Essays on Iron and Steel*, pp. 110-111.

84 *Pennsylvania Mercury*, August 10, 1787.

85 Israel Acrelius, *History of New Sweden*, p. 166.

86 M. Jars, *Voyages Metallurgiques*, pp. 257-258 ; Samuel Smiles, *Industrial Biography*, p. 104.

87 *Pennsylvania Gazette*, August 5, 1789 ; *Pennsylvania Packet (Claypoole's American Daily Advertiser)*, September 4, 1797 ; *Cramer's Pittsburgh Magazine Almanack*, 1812.

88 *Pennsylvania Packet (Claypoole's American Daily Advertiser)*, September 4, 1797, etc.

89 P. W. Flower, *Short History of the Trade in Tin*, pp. 45, 93 ff.

90 *Cazenove Journal*, pp. 6 ff ; *Pennsylvania Packet* (Dunlap), May 14, 1791.

91 *Pennsylvania Packet (Dunlap and Claypoole)*, May 3, 1795.

92 *Pennsylvania Packet*, October 7, 1788.

93 *Pennsylvanische Staatsbote*, July 21, 1775.

94 James Mease, *Archives of Knowledge*, II, 380.

95 *Philadelphia Federal Gazette*, September 1, 1796.

96 *Ibid.* November 29, 1793, April 21, 1795.

97 G. Agricola, *De Re Metallica*, pp. 423 ff.

98 R. A. F. de Reaumur, *Art de Convertir le Fer en Acier.*

99 John Percy, *Metallurgy*, pp. 772-773.

100 Israel Acrelius, *History of New Sweden*, p. 164 ; Joseph Scott, *Gazetteer of the United States*, 1795, Article on "Pennsylvania."

CHAPTER V

IMPROVEMENTS AND INVENTIONS

D URING the first part of the eighteenth century, as the iron industry was being established and was beginning to grow, few improvements or inventions could be expected. The technological difficulties that had to be solved by the first generation of ironmasters, who as a whole had not much experience in this line of industry, left little time for experimentation or for the consideration of untried plans. It has been pointed out the American iron manufacture in its various branches was patterned after that of the mother country of the same period. This was true in regard to blast furnace production, including the type of furnace, blast, fuel and methods of casting. It was also generally true of the forging of iron as well as the manufacture of iron in its secondary branches.

About the time of the Revolution a change was made in a new type of blowing apparatus. This was the substitution at a number of blast furnaces and forges of blowing cylinders or "blowing tubs" for the large wood and leather bellows in the production of the blast. Such cylinders were the invention of the English inventors, James Knight and John Smeaton. This new means of producing blast came into general use in England after 1762.[1] By the time of the struggle between the colonies and the mother country similar blast-producing cylinders were in use in Pennsylvania and in other iron-producing regions in America, especially at the forges. These were copied after English models. Attempts were made at Cornwall Furnace to produce such cylinders of cast iron.[2] Evidently, little success was obtained, for the vertical wooden blast cylinders continued to be used for a long time as is evidenced in the accounts of such appliances well into the nineteenth century.

As time went on greater varieties of castings were made, as for example can be seen in the development of different types of stoves. In comparing stoveplates made during different periods in the eighteenth century, progress can be noted throughout the century, not only in design but in the quality of workmanship. Cast iron came to be improved in quality by experimentation in mixing different ores. Ironmasters, also began using iron recovered from cinder banks because it was discovered that mixed with ores it produced a much tougher iron than heretofore. The unfortunate Peter Hasenclever in his enterprises in New Jersey and New York was largely responsible for the widespread adoption of this plan. In 1765,

Stove Plate

his partners wrote him from London regarding such iron: "We can now, with the utmost pleasure and satisfaction, acquaint you, that it is universally allowed by the trade to be the best drawn Iron, by far, that ever made its appearance in the London Market from America; it has been tried and found of exceeding good quality." Hasenclever replied proudly: "There is, perhaps, no better Iron in the world than this cinder iron, particularly for the use of wagon tire, plough shares and implements for husbandry; it is tough and almost as hard as steel."[3] For certain purposes, therefore, cinder iron had its uses. During this period stamping mills were established at the ironworks to crush the furnace cinders and slag in order to secure more easily the metallic iron from the slag.

Quite early in the history of the province the Germans introduced stoves and from the beginning of the industry, stoves and firebacks were cast at the blast furnaces. In most eighteenth century homes, however, stoves could not be found, the wide, open fireplaces being used for cooking and heating. That there was a market for stoves, though, is clearly evidenced in the sales recorded in the ledgers of the ironmasters and also in the improvements made from time to time in different types of cooking and heating appliances.

One of the most important changes that took place in heating stoves was the invention by Franklin in 1742 of his fireplace. It was described by an European traveler as "a sort of iron affair, half stove, half fire-place a longish rectangular apparatus made of cast iron plates and stands of far from the wall the front being open, in every respect a detached moveable fire-place."[4] In his autobiography, Franklin states that he presented a model of his invention to his friend, Robert Grace, of Warwick Furnace, who found the casting of the plates for the fireplace profitable.[5] Franklin himself sold a number of these stoves in Philadelphia.[6] But he refused to secure a patent for his invention, stating that he believed it should be used for the benefit of humanity.[7] Others took his ideas and manufactured fireplaces that became known as Franklin stoves. While in London in 1765, Franklin wrote to Hugh Roberts asking him to send two of the stoves cast years before by Robert Grace. To this Roberts replied:

> Thy request of procuring 2 Pens[a] Fire Places cast when Robert Grace's Moulds were good, is a little uncertain which of the sorts thou intended, whether the first impression such as Psyng and I had or that with the Sun in front and Air Box, both of which are much out of use, and tho' many have been laid aside I find on enquiry yt some parts of the plates have been apply'd for Backs or hearths of Chimneys or other Jobs, that I have not found a second hand one compleat. . . .[8]

By this time many improvements had been made in the stove and it was known even in Europe. Many were reaping the benefit of Franklin's principles.

During the closing years of the century, spurred by the premiums offered by the American Philosophical Society, a number of inventors were working on improvements to stoves, especially to the Franklin stove.[9] One inventor, in 1796, pointed out that most homes possessed no stoves, but that open fire-places were in general use, although they were being made

Franklin Open Fireplace

smaller in Philadelphia and in other well-populated sections of the country on account of the increasing scarcity of fuel. In his communication he stated:

> The close-iron-stove was invented to save fuel by casting heat with the least possible loss. But it is far from solving the question in full latitude. By being unattended with a free circulation of air it is insalubrious. It is also expensive and admits not of cooking. To remedy some of

the inconveniences of the close-stove, particularly the insalubrity the immortal Franklin invented his open-iron-stove. This by its openness
gives a free circulation of air and by its projec-
tion into the room a back-plate with a space
behind it communicating to the room causes but
a little waste of heat produced by the fuel. But
at the same time it increases the consumption of
fuel beyond the close stove. It is also expensive
and does not admit of cooking. This stove is an
invaluable acquisition to the richer part of the
world, but the poor can never enjoy it. . . .[10]

The writer, therefore, planned to incorporate the advantages
of the Franklin stove into what he called a stone stove, a
creation of his own mind.

While stove-plates and a variety of iron wares were cast at
the blast furnaces the better grade of castings were made at
air furnaces. Among those who carried on such furnaces in
Philadelphia was John Nancarrow, a man of inventive genius.
Nancarrow devised many improvements in the methods of
casting and also in articles that were cast. Newly invented
boxes for carriage wheels were made in Philadelphia in 1785,
and during this period many improvements were made in tools
and machinery made of cast iron. The cylinders required for
steam engines during this experimental age were sometimes
cast at the blast furnaces, but more often at the air furnaces.

Experiments in perfecting the steam engine were made during the last quarter of the century at the same time that Watt,
Boulton and others in England were pushing their inventions
from an experimental to a practical stage. Before the Revolution, two steam engines were introduced into New England,
but without much success. About the same time one was installed at the Schuyler Copper Mines on the Passaic River,
in New Jersey. The principal parts of this engine were imported from England and an engineer, Hornblower, was sent
with it to superintend its construction. Writing of it at the
beginning of the nineteenth century, Benjamin H. Latrobe
stated that in spite of its imperfect construction and the faulty
boring of its cylinder, it had drained the mine effectually for
thirty years.[11]

Among the first experiments in engine building in America was one made in 1773, by Christopher Colles, an ingenious Irishman. This learned mathematician and engineer delivered lectures on "pneumatics, hydrostatics and hydraulics" before the members of the American Philosophical Society. He built a steam engine designed to pump water at a distillery, but the experiment was not successful.[12] Another attempt was made in this country in 1775, when Sharp and Curtenius built a steam

Fireplace Cast at Hopewell Furnace

engine for the water works at New York. This was adjudged at the time to have been well executed.[13]

The difficulties of the Revolution retarded the progress of invention. In the years that followed the peace, many turned their attention to the use of steam for propelling boats. La-trobe wrote that "a sort of mania began to prevail for im-pelling boats by steam engines."[14] Even the venerable Franklin was much interested. It was during this period, too, that the unfortunate John Fitch received from the legislature the ex-clusive right to his invention of navigating vessels by steam.[15]

The story of Fitch's struggles, his steamboat that plied the Delaware for a time, his failure and death are well known and need not be recounted here.

Among others who were experimenting with steam power were Arthur Donaldson, Doctor Kinsey, Henry Voight and James Rumsey. Many improvements in the boiler, pipe, condenser and other machinery by Fitch, Voight, Thornton, Hall, and Evans led to the development of the engine which operated Fitch's boat on the Delaware during the summer of 1790. Latrobe, surveying the attempts to perfect the steamboat, wrote early in the nineteenth century: "Nothing in the success of any of these experiments appeared to be sufficient compensation for the expense and the extreme inconvenience of the steam engine in the vessel."[16] He pointed out that the chief objections to the steamboat included the weight of the engine and fuel, the large space occupied by both, the vibrations of the engine that often racked the vessel, the irregularity of the boat's motion, the expense of steam power, and the difficulties connected with securing paddles of the right weight for if too light they would break and if strongly made they were too heavy. Fitch tried to overcome the latter objection by his experiments with a screw propeller.[17]

Attempts were also made to apply steam power to land carriages. Oliver Evans received patents in several states to the right to operate his "Columbian," but he failed to interest the Lancaster Turnpike Company in his invention designed to draw passengers and freight over the new highway. In 1804 Evans produced his strange "Orukter Amphibolos," designed to travel on land and water, and in the years that followed he applied his ideas in the construction of new "Columbian" engines.[18]

At the opening of the nineteenth century there were few steam engines engaged in practical uses in the United States. There was one owned by the Manhattan Water Company, New York, built on the principle of the Boulton and Watt double engine. Another, also in New York, used at a lumber mill, was built by Nicholas Roosevelt at the Soho Works, New Jersey. In 1799 the city fathers of Philadelphia decided to lay down water pipes on the principal streets for the convenience of householders and as a protection against fire. The plan of the engines to pump the water was executed by Latrobe. Two low pressure engines, built at the Soho Works after the Boulton and Watt plan drove, not only the pumps at the water works, but also a rolling and slitting mill. Another steam engine was used at a factory at Boston, and Oliver Evans was operating a small one used to furnish power for grinding plaster of

Ten Plate Stove

Paris.[19] Many parts for steam engines were cast at such furnaces as Warwick and especially at the air furnaces at Philadelphia. Soon after the opening of the nineteenth century, Oliver Evans began to build steam engines for others at Philadelphia and may, therefore, be considered to be the first regular builder of steam engines in the United States.

The application of cast iron to the building of bridges became a reality first in England, although one of the early plans for bridging rivers with iron was worked out in Pennsylvania. Thomas Paine, the Englishman who had played so prominent a part in the American Revolution, turned his attention to engineering and in 1786 presented Franklin with the model of a bridge designed to span the Schuylkill River. Paine proposed to build an iron bridge 400 feet in length—a single arch with thirteen ribs representing the strength and unity of the states of the new republic. Before the project was accomplished Paine left for England and took out a patent there for his invention.[20]

At Masborough, an arch of ninety feet was built to span the river Don after Paine's plans. Another bridge, a little longer, was cast and was placed on exhibition at Paddington. At the outbreak of the French Revolution, Paine turned his attention from bridge-building and hastened to Paris to join the "Friends of Man." From the modification of designs he left behind him, another bridge was built over the River Wear at Sunderland in 1796. In the United States, little attention was paid to building bridges of cast iron, although some bridges of iron chain, such as the one across Jacob's Creek in western Pennsylvania appeared at the turn of the century. Built by Judge James Finley of Fayette county, it claimed consideration and many chain bridges were built on Finley's plan in different parts of the country.[21]

During the closing decades of the century many attempts were made to improve the quality of blister steel produced in the country as well as to expand its manufacture. Whitehead Humphreys, assisted by the legislature made many "discoveries in the art of converting bar iron into steel," and made improvements in the trial and error method of producing the metal.[22] Henry Voight, also discovered an improved method of making steel which was claimed to be better than imported cast-steel, especially for use in the best grade of knives and razors.[23]

During the last three decades of the century many attempts were made to produce iron wire, but without much success. Machines worked by water power were set up. An early attempt to draw wire in Philadelphia was made in 1775.[24] Four years later Nicholas Garrison, Valentine Eckert and Henry Voight erected a wire factory, but the experiment was not successful.[25] Not until after the close of the century when Eichbaum built a wire mill, run by steam power in Pittsburgh with financial aid from the state legislature were the attempts to draw wire successful.[26]

In manufacturing iron wire to be used for making cards for combing cotton, wool and flax, more success was attained. In 1777-1778, Oliver Evans invented machinery for manufacturing wire from bar iron and for making cards from the wire. He proposed to establish his factory under state patronage, but being unsuccessful in this he sold his secret to a company at Wilmington for a relatively small sum. Later he devised a plan for pricking the leather, and for cutting, bending and setting the teeth. Because of the "impoverishing result of his previous discovery," he never carried this plan into execution.[27] It was worked out later by Amos Whittemore in Massachusetts. However, by 1797 there were three manufacturers of iron wire cards in Philadelphia, turning out thousands of dozens of cards annually.[28] Factories for producing cards had also been established in Boston and in other sections of the country.

Improvements were also made in the making of nails. In 1788 John Fitch invented a machine for rolling and cutting nails. Henry Voight accused him of theft of the invention. Finally, however, the two men agreed to work together and backed financially by Richard Wells and Thomas Clifford, Jr., plans were made to sell rights to use the machine to English industrialists.[29] The records are silent, however, as to whether any progress was made beyond the experimental stage. One year later, Samuel Briggs, who also experimented with steam power, petitioned the state legislature and the federal Congress regarding a machine for making nails, screws and gimlets. A model of his invention in a sealed box was deposited with the governor of Pennsylvania. Together with his son, he received the first patent issued under the general patent laws of the United States for nail making machinery.[30] Other patents were granted for similar machinery before the end of the century.

Among other inventions were plans to perfect the drawing of bar iron into round shapes. This was accomplished as early as 1783 at the works of Mark Bird and James Wilson at the falls of the Delaware by hammering the heated bars under concave rolls.[31] In 1796, Clemens Rentgen discovered a new method of making round iron[32] and in 1810 he invented rolls which in some ways were similar to those of Cort in England. There were a number of minor inventions toward the close of the century such as that of Emanuel Bantling of a "tube bellows" for blacksmiths.

FOOTNOTES

[1] T. S. Ashton, *Iron and Steel in the Industrial Revolution*, p. 37; H. Scrivenor, *History of the Iron Trade*, p. 82.

[2] S. G. Hermelin, *Report About the Mines in the United States of America, 1783*, p. 63.

[3] *The Remarkable Case of Peter Hasenclever*, pp. 10, 80.

[4] J. D. Schoepf, *Travels in the Confederation*, I, 60-61.

[5] Benjamin Franklin, *Autobiography*, p. 139.

[6] B. Potts MSS., X, Popadickon, 1750, pp. 215, 219.

[7] Benjamin Franklin, *Autobiography*, pp. 139-140.

[8] Hugh Roberts to Benjamin Franklin, 27, 11 mo. 1765, Charles Morton Smith MSS., Document 46.

[9] MSS. Communications to the American Philosophical Society: Navigation, Manufactures, Agriculture, and Economics, Documents 11, 18, 20. Franklin himself in 1786, invented a new type of stove for Burning Pitcoal and Consuming all its Smoke." American Philosophical Society, *Transactions*, II, 57-73.

[10] "An Essay on Warming Rooms," MSS. Communications to the American Philosophical Society: Navigation, Manufactures, Agriculture and Economics, Document 18.

[11] "Report of Benjamin Henry Latrobe," MSS. Communications to the American Philosophical Society: Mechanics, Machinery and Engineering, Document 25.

[12] Thompson Westcott, *Life of John Fitch*, pp. 153-154.

[13] *Rivington's New York Gazetteer*, February 16, 1775.

[14] "Report of Benjamin Henry Latrobe," MSS. Communications to the American Philosophical Society: Mechanics, Machinery and Engineering, Document 25.

[15] Minutes of the General Assembly 1786, 10th Assembly, 2nd Session, pp. 199; 205-206, 230; 11th Assembly, 1st Session, pp. 40, 44, 99-100; *Pennsylvania Statutes at Large*, XII, 1.286.

[16] "Report of Benjamin Henry Latrobe," MSS. Communications to the American Philosophical Society: Mechanics, Machinery and Engineering, Document 25.

[17] Thompson Westcott, *Life of John Fitch*, pp. 363-364.

[18] G. and D. Bathe, *Oliver Evans*, pp. 6ff.

[19] "Report of Benjamin Henry Latrobe," MSS. Communications to the American Philosophical Society: Mechanics, Machinery and Engineering, Document 25.

[20] Specifications of Patent, No. 1667, (1788).

[21] J. M. Swank, *Progressive Pennsylvania*, p. 248.

[22] *Votes of the House of Representatives*, 1770, VI, p. 200, 208; *Pennsylvania Chronicle*, October 24, 1772; *Pennsylvania Packet*, October 26, 1772; *Pennsylvania Packet Supplement;* January 18, 1773.

[23] Thompson Westcott, *Life of John Fitch*, pp. 402-404.

[24] *Pennsylvanishe Staatsbote*, July 21, 1775.

[25] J. T. Scharf and Thompson Westcott, *History of Philadelphia*, III, 2260.

[26] James P. Mease, *Archives of Useful Knowledge*, II, 380.

[27] Patrick N. I. Elisha (Oliver Evans) *Patent Right Oppression Exposed*, pp. 20-24.

[28] J. T. Scharf and Thompson Westcott, *History of Philadelphia*, III, 2260.

[29] Clifford Papers, 1788-1790, IX, Documents 205, 206, 207, 209, 234, 274, 284, *passim.*

FOOTNOTES—Concluded

[30] J. L. Bishop, *History of American Manufactures*, I, 571. A machine for producing cut iron nails was established in western Pennsylvania before the end of the century, *Pittsburgh Gazette*, July 22, 1797.

[31] S. G. Hermelin, *Report About the Mines in the United States of America, 1783*, pp. 74-75.

[32] F. Sheeder, "East Vincent Township," *Pennsylvania Magazine of History and Biography*, XXXIV, 376. *The Aurora*, December 3, 1796 advertised Clements Rentgen's new machine for drawing round iron ¼" to 2" in diameter or larger.

CHAPTER VI

THE WORKERS

THE labor system that developed in the American colonies was based upon English labor institutions of the seventeenth century. Many modifications, however, were made because of conditions imposed by a large country that had a small and scattered, but growing population. In the New World, land was plentiful and cheap, whereas labor was usually scarce and expensive. As a result, English standards and practices were modified to conform to a different social and economic environment. The status and conditions of labor were molded to fit into the economy of a frontier country. This was evident in fewer restrictions and regulations both by law and custom in America in comparison with the mother country. Then again, servile labor occupied a far more important place in the American colonies than in England. It was natural that the continual scarcity of labor should result in the extensive use of indentured labor and even slaves in industry.

A few factors in the background of colonial industrial institutions should be noted in order to understand labor conditions in America. At the close of the Middle Ages and even earlier the production of iron in Europe, differing from other industries, was necessarily a capitalistic enterprise in the sense that workers were brought together in a plant to labor, the employer furnishing the equipment and raw materials and also marketing the finished products; the workers supplied the labor and received wages in return for it. The progress made throughout the centuries in the iron industry in general was not achieved in the same manner as in other industries, such as textile manufacturers, through changes in a craft or guild system, to a domestic system where work was done for the employer in the home, and then to the factory system which was brought about by the Industrial Revolution. Instead, the organization of the iron industry remained the same, the changes that took place resulting from the expansion of the industry due to the opening up of new sources of raw materials, as well as to the development of markets. Inventions and technological improvements, of course, inevitably accompanied expansion and contributed to it, often being responsible for it.

As a result of such organization, ironworkers in England and in other countries were not subject to the close restrictions imposed by the guilds as in other industries, although through legislation and custom certain regulations were enforced. A period of apprenticeship, with definite duties and obligations for master and apprentice, for example, was required, especially in the forges and in various branches of secondary iron manufacture, such as nail-making, lock-making and blacksmithing. In the New World, however, as manufactures developed, restrictions were much less harsh in their terms and opportunities for the worker, therefore, were greater. Class lines, especially within the trades, were less rigid; the conditions of work were freer; and the opportunities for advancement were much greater.

The problem of obtaining a sufficient supply of labor confronted ironmasters and all employers continually.[1] It was especially difficult to secure skilled workers for furnace and forge. A traveler in the decade of the twenties pointed out that iron manufacture was held back to a considerable extent in the growing colony of Pennsylvania because of the "want of furnace men, smiths and charcoal burners."[2] A few ironmasters like Samuel Nutt, William Branson and others made occasional voyages to Europe in order to supply this need, bringing back with them skilled and unskilled laborers.[3] The number induced to try their fortunes in the New World by the individual ironmasters was not large, usually only a few who became indentured servants for a specified number of years in return for their passage across the ocean.

Ironmasters, too, like other employers, often secured workers from among newly-arrived immigrants from different parts of Europe. When vessels reached Philadelphia, captains or merchants would advertise the sale of indentured servants —men and women who had indentured themselves in order to pay their passage. Occasionally skilled workers were found among these, although most of them were unskilled. Still another means of securing labor from the Old World was through the interest and influence of relatives and friends who persuaded workers to leave their homes for the colonies. For example, in 1764, Thomas Penn, who at that time was residing in England sent a number of experienced colliers or charcoal burners to the province for the special purpose of "teaching those employed at the Iron Works to make the coal [charcoal] with a much less quantity of wood than they had hitherto consumed."[4] Thus, in various ways workers were secured from the Old World.

The methods used by Peter Hasenclever in securing labor for his brief and unsuccessful industrial experiments in New Jersey and New York were not at all typical of eighteenth century America. In fact, his entire scheme was one of a few isolated attempts at what might be called large scale industry in the colonies. The project was planned in England by Hasenclever, a Prussian, who moved to London at the close of the Seven Years' War and became a naturalized Englishman. After establishing the mercantile house of Hasenclever, Seton and Crofts in London, he became convinced of the opportunities afforded by the natural resources of the mainland colonies. With real organizing genius, he planned one of the most elaborate and extensive industrial enterprises of colonial America, "The American Iron Company," or as it was often called, "The London Company."

Hasenclever planned to engage not only in ironmaking, but also in the manufacture of potash and in the production of hemp, flax and madder. He purchased more than 50,000 acres of land in New Jersey and New York, and later petitioned the British government for large tracts of lands in Canada, which were not granted him. Within a short time he had spent a great amount of money on ironworks at Charlotteburg, Ringwood and Long Pond in New Jersey and at Cortlandt in New York. In 1765 and 1766 he built four blast furnaces and seven forges as well as houses, barns, grist mills, saw mills, blacksmith shops, bridges, canal, dam and reservoirs.

For his enterprises, Hasenclever transported 535 men and their families from Germany as well as a number from England. Among this large group of workmen were miners, founders, forgemen, colliers, carpenters, masons, and laborers. Many, if not all, of these workmen were engaged by contract for a period of years and their status was not exactly that of indentured servants, for the contracts consummated in Germany provided for wage-payments. Certain of the workers—especially unskilled laborers—received a daily wage; the skilled artisans were paid on a piece-work or tonnage basis. This is one of the earliest illustrations of the foreign contract labor plan, which a century later developed into a system with its attendant evils until Congress prohibited the importation of alien contract laborers in 1885.

The industrial experiments of Hasenclever were not successful. Difficulties with his workmen chiefly over the question of wages which many of them believed had been fixed too low in the contracts, the desertion of many skilled workers as is evidenced in the advertisements which appeared in the newspapers of the times, the dissatisfaction among members of the company in England because the unfortunate manager had exceeded his instructions in spending more than £54,000 instead of the £40,000 pledged and also because profits were not forthcoming, and finally the bankruptcy, in 1770, of the London firm of Hasenclever, Seton and Crofts resulted in the recall of Hasenclever and in the downfall of The American Iron Company. During the years that followed there was much litigation over the properties of the company and the various works passed into the hands of Americans.[5] As the projects of Hasenclever failed, many of the workers, skilled and unskilled, dispersed to other regions, a number of them finding employment in the ironworks of Pennsylvania about the time the war clouds were gathering on the eve of the Revolution.

As in neighboring sections of the country where ironmaking was carried on in the eighteenth century, the workers in the Pennsylvania iron industry constituted a heterogeneous group. English, Welsh, Scotch, Irish, Scotch-Irish, German, native-born American, Negro, and occasionally Indian workmen, must be included in a survey of those who produced iron or who assisted directly or indirectly in its manufacture during the period under discussion. The workers may be classified in a general manner as follows: (1) free labor; (2) indentured servants or redemptioners; (3) Negro slaves and freed Negroes, and (4) journeymen and apprentices who were employed especially in the production of tin plate, nails and secondary iron manufactures in Philadelphia and in the boroughs where such works were found. Child labor found small place in iron manufacture, although at the end of the century children were used to produce iron wire cards used for combing cotton, wool and flax.

Even though skilled labor was always scarce and sometimes mobile, there were occasional instances of families of iron-workers who remained for generations on the same plantation. The furnace ledgers and the correspondence of early iron-masters mention the names of some of the workmen over long periods of time, and families in a few cases can be traced for two or three generations from these records.[6] A number of the ironworkers of the Windsor Forges in Lancaster county, present an excellent example of this kind, for the descendants of some of the first workmen at these forges were employed at the time the career of the ironworks came to an end, a century and a half after their establishment during the colonial period.[7]

The wars of the eighteenth century added to the difficulties of the labor supply. When, in 1740, the governor of Pennsylvania called for volunteers to attack the Spanish West Indies,[8] several hundred indentured servants responded by enlisting. In June, 1741, Anna Nutt and Company, the owners of Coventry and Warwick Iron Works, complained to the Assembly that ten of their indentured servants had enlisted, many of them being colliers who could not be replaced. The ironworks had been forced to shut down, with a resultant loss to the company of several hundred pounds. To offset the loss through enlistment, the Assembly adopted the policy of compensating out of the public funds the employers who had suffered losses in this manner.[9]

Again, during the French and Indian War the same problem appeared. Pennsylvania's lack of interest and her half-hearted support in the war, the opening bloody scenes of which were unfolded within her boundaries, have long been familiar. While this attitude was largely due to the policy of the peace-loving Quakers, it also resulted from the dissatisfaction brought about by the enlistment of indentured servants and other laborers at a time when labor was extremely scarce. This is clearly evident in the debates which preceded the law passed in 1755 by the Assembly for regulating enlistments. Under the act minors, indentured servants, or indentured apprentices should not be permitted to enlist without the written consent of parents or masters. The act was promptly disallowed by the English government and the governor, the Assembly and others concerned were directed to "take notice and govern themselves accordingly." Other attempts were made by the Assembly to regulate the enlistments of such servants in spite of the fact that Parliament in 1756 enacted that British officers could enlist any indentured servant and that such officers were not to be bound by colonial law, custom or usage in this respect. However, property rights were recognized in that a master who objected within six months after the enlistment of a servant could recover him or else be compensated as two justices of the peace should decide. As masters had no satisfactory recourse against the British government, the Pennsylvania Assembly once again compensated masters for the loss of labor during the war.[10]

During the Revolution when the furnaces of southeastern Pennsylvania were busy casting cannon, shot and ironwares for the Continental armies, the workmen at these ironworks were not permitted to leave their employment, or to march with the militia without the permission of the Council of Safety.[11] So great was the scarcity of labor that the ironmasters of Chester, Lancaster and Berks counties were permitted to use prisoners of war at their works.[12] Under this authorization, Hessians were employed at Elizabeth Furnace, Mary Ann Furnace, Durham Iron Works, Charming Forge and at other ironworks.[13] Twenty-two Hessian prisoners were employed at Elizabeth Furnace for which the Continental Congress received thirty-two to forty-five shillings a month for each of them, the amount being paid in iron.[14] At Charming Forge, thirty-four Hessian prisoners were employed to cut a channel through a bed of rock to supply the slitting mill—one of the types of works proscribed by the British government—with water power.[15] For their services, George Ege, the ironmaster, paid the government £1020 in iron.

The skilled workers on the iron plantations were usually English, Welsh, Irish, and German,[16] although quite often freed Negroes and Negro slaves filled such positions at the forges. The managers, founders, keepers, guttermen, and potters at the furnaces were men of experience and understanding as well as skill. Likewise, the "finers," chafery men and hammer-men at the forges, who heated the metal in hearths and drew out the bars to given sizes under ponderous hammers, had to be men of some intelligence, who had spent many years learning their occupations. The skilled workers at the forges, slitting mills, plating mills, nail works and tin plate works were

Hammer Heads and Wedge
The large hammer head, used at Poole Forge, Lancaster County, weighs about 500 pounds.

journeymen who had served an apprenticeship in a system not unlike, but not as restrictive as the system in European countries.[17] Blacksmiths occupied a very important position for as a group they were skilled artisans and most of the secondary iron products such as axes, hoes, shovels, chains, etc., were made by them. Among the skilled workers must also be classified the gunmiths and locksmiths, who worked at their crafts alone or with the aid of a journeyman or two, or a few apprentices.[18]

The indentured servants or redemptioners were generally brought from Great Britain or Germany, but in some instances

Negroes and mulattoes served as indentured servants.[19] Occasionally skilled workers, unable to pay their way across the ocean, became indentured to the ironmaster for a period of years.[20] While a few indentured servants worked as "finers" or hammermen in the forges, most of this class of workers were common laborers. They worked around the mines "broke" ore on the "bank," hauled the ore, limestone flux and fuel to the furnace top, drove the teams, repaired fences, dug ditches, plowed and worked on the plantation, and performed all kinds of necessary menial labor. Quite frequently, servants who were restless, discontented or adventuresome ran away from the plantations, and many advertisements seeking their return appear in the newspapers of the time.[21] It was usually difficult to find and to secure the return of such fugitives. Runaways occasionally fled from one iron plantation to another[22] Such instances, however, were rare, for in seeking employment as ironworkers they placed themselves in a position that could have led easily to their discovery. The price paid for white indentured servants was not high, but varied from time to time. The trend was in the direction of a gradual increase through the century. About the middle of the century indentured servants were usually sold for £14 for four years' service or £18 to £20 for a term of five years.[23]

Negroes were used in the ironworks from the early establishment of the Pennsylvania industry.[24] In 1727, the shortage of labor was so acute that the ironmasters in the colony petitioned the Assembly for permission to import Negroes free of duty to labor at their works. The petition set forth that the "difficulties of getting laborers and their excessive wages are a great discouragement and hindrance" to the development of iron manufacture.[25] A bill permitting Negroes imported into the colony for the express purpose of laboring at ironworks to enter duty free, failed by the deciding vote of the Speaker. Two years later, however, the duty of £5 on each Negro brought into the province, was reduced to £2.[26] This was a concession to the industrialists. A few years later, however, in accord with British policy notice was given to Deputy Governor Gordon that he should not give assent to any law imposing duties on Negroes imported into Pennsylvania because the duty operated "to the great discouragement of Merchants trading thither from the coast of Africa."

While Negro slaves and freed Negroes usually worked at menial tasks, yet at many ironworks they were skilled workmen. While few skilled Negro workers were found at the blast furnaces many were employed at the forges. At Green Lane, Durham, Martic, Pine, New Pine, Mount Joy (Valley), Charming, Pottsgrove and others,[27] they performed the skilled tasks of refining and drawing iron into bars.

The cost of slaves employed at the ironworks varied. In 1728, the Coventry ironmasters bought two slaves for £74;[28] in 1775 £150 was paid for "Negro Dick and his wife" at Pottsgrove,[29] while about 1750 "Cato and Cudgo" cost £64-10-0 and £54 respectively.[30] Acrelius states that at the middle of the century slaves cost from £30 to £40.[31] Perhaps £50 would be closer to the actual average for full grown Negroes at that period. When slaves became skilled workers, they were very profitable, for the ironmaster had only to feed and clothe them. They were treated very kindly and considerately in eighteenth century Pennsylvania as may be seen from the records of small sums of money and other gifts given them, especially at the holiday seasons.[32] As was the case with indentured servants, there were occasionally some who became restless and sought their freedom by running away.[33] Here again it was usually difficult to effect the capture and return the fugitives.

In contrast to English iron manufacture of the same period, women had little or no part in the production of iron. In England the wives and even the children of the regular workers were employed for such purposes as preparing the ore, picking pieces of iron and charcoal from the furnace cinders to be used again, and repairing charcoal baskets and sacks. Many women, however, usually the wives of regular ironworkers, labored on the plantations at certain periods and therefore may be classified as casual farm workers. They made hay, pulled turnips, reaped the grain, and in fact, did much of the harvesting as is set forth in hundreds of items in the account books of the period. Not only did they work on the fields, but they spun thread and wove cloth for the ironmaster and his family. They were credited, like their husbands, for the work done each day, in the large furnace and forge ledgers, the rate varying according to the tasks performed.[34]

The few Indians occasionally employed as common laborers at Colebrookdale and at some of the other ironworks might be mentioned in passing. These workers never could be depended upon. Too frequently they heard the "call of the wild" and left the plantation to roam afar. Sometimes they returned bearing

Refinery Forge Hammer

gifts of pelts and furs for the ironmaster, desiring to be employed again, or to receive credit at his store. During the first half of the century especially, there was much barter between the ironmasters and the Indians as is evidenced in the account books and correspondence that remain.[35]

Another group of workers in Pennsylvania iron industry that must be considered were the apprentices and journeymen, employed for the most part in secondary iron manufactures. Boys were employed at many forges as helpers and learned the trade of forging. At the slitting mills, plating forges (tilt hammers) nail works, and tin plate works, journeymen and apprentices were employed. From twenty to thirty worked in each of the nail and tin plate works which were located in Philadelphia, and in the boroughs and towns.[36]

The first general law for the regulation of apprentices within the province, which definitely established the apprentice system, was passed in 1763.[37] Under this act, indentured papers had to be drawn up, males were bound until they were twenty-one years of age, while females served until they were eighteen. If the master abused his apprentice or failed to live up to the covenants of the indenture, the apprentice might apply to any justice of the peace, who was authorized to settle the grievance. If this could not be done, the master was bound over to appear at the next county court of quarter session. Likewise, if the apprentice failed to do his part, the master could seek redress by applying to the nearest justice of the peace. If no agreement could be reached, the apprentice would have to put up surety to appear at the next session of the county court, or else be committed to jail. In the case of runaway apprentices, any justice of the peace could issue a warrant to any constable within the county or city to catch and apprehend such apprentice, and have him appear before any justice of the county. The apprentice had to consent to return, find surety to appear at the next session of the court in the county where his master resided, or be committed to jail. Anyone harboring a runaway apprentice was required to pay the master twenty shillings for every day the apprentice was harbored. This statute was re-enacted in 1770.[38]

The indentured, drawn up before a magistrate, provided that the boy, who was usually from ten to fourteen years of age, was apprenticed until of age. The document specified his duties and obligations to his master.[39] The latter, on his part, promised to clothe and feed the boy, teach him the trade under the direction of a competent journeyman, and have him instructed in reading, writing and arithmetic. When the apprentice became of age, he was to receive a new suit of clothes and a small sum of money to buy a set of tools.[40] The apprentice then stepped into the ranks of the journeymen.

The relations between the ironmasters and the workers varied to a great degree. Negro slaves and indentured servants, who received no wages, have already been mentioned. With all other classes of workmen written contracts were frequently made. More often, the agreement was simply oral. In the latter case, an entry in the ledger usually, but not always, set forth the date and the terms of the contract.

Written agreements, sometimes duly witnessed, were made by the ironmaster, not only with the more highly skilled workers such as founders, keepers, colliers, forgemen and finers. but frequently with wood-cutters, stockers, and carters. Agreements were at times written in the ledgers and signed by both parties. Some occasionally contained unusual provisions as may be seen in the following:

> Agreement with John Shaw July 23rd. 1761 to stock [with charcoal] the upper Forge, and at any time to assist in stocking at any of the other two forges when he has not stocking to do at the said upper forge. The said Shaw is to be paid for the faithful performance of the above agreement eighteen pounds and a pair of shoes, and if he does not get drunk above once in three months, a pair of stockings and his diet.

> Oct. 6th. 1762 Agreement between John Boyer and Patton and Bird to drive team one whole year from the date hereof, for the due performance of which said Patton promises to pay him the sum of twenty-six pounds and two pair of shoes.[41]

Other agreements made between employer and employe include, besides a certain agreed wage, "meat, drink, washing and lodging."[42]

The wages paid the ironworkers did not vary to a considerable extent in the different regions, and throughout the century there was but little change, although there was a tendency to a slight gradual rise. Wages were nominally higher at the close of the century than they had been in the opening years

of the industry. This was due to the general upward movement in the prices of commodities which, however, was not
great, together with the increased demand for iron for the
manufacture of secondary iron products. The breaking away
from the mother country gave an impetus to American manufactures, while the European wars of the last decade of the
century aided the progress of manufactures by placing obstacles in the shipment of wares from England.

Each furnace had a manager, whose duties often included
the necessary bookkeeping. The pay of managers varied from
£70 to £120 a year. The highest paid of the furnace workers
were the founders.[43] They were usually paid by the ton,
although occasionally in the early days of the industry, they
were paid partly by the day.[44] The rate paid founders, or
"smelters" as they were sometimes called, before the middle
of the century was two shillings and six pence per ton for
casting pigs, and about twenty shillings per ton for country
or small castings.[45] In 1783 and 1786, the same rate was paid,
the two founders receiving together five shillings per ton for
pigs.[46] An average furnace, during the latter decades of the
century, turning out twenty-five tons per week would net the
two founders working alternate shifts about £6, or £3 per week
each. This would amount to about £12 per month each. At
some furnaces there was a head founder and a second founder.
The former generally averaged £15 per month and the latter
£10 for the same period.

When the founder laded out the metal to be used by the potters or molders who made country castings, an operation that
required the frequent stopping of the furnace, he received
twenty shillings a ton as the operation took a much longer
time than casting pig iron. When the founders themselves
made open castings, such as stoveplates, fireplaces, pots, flat-
irons and wheel bushings working without the aid of the potters, they received forty shillings a ton, but out of this they
paid two keepers.[47] This higher rate resulted in a slight increase in the wages of the founders. By way of comparison it
might be pointed out that free common labor at the ironworks
received from £2 to £4 per month or when food and lodging
were included £1 to £2 per month.[48] It should be noted that the
furnaces generally remained in blast only about eight or nine
months in the year. During the winter season, the wood was

cut, charcoal was made, dams repaired, bridges and roads built, as the weather permitted. Founders who turned to these tasks while the furnace was out of blast received a daily wage.

The ironmasters often attempted to hold the founders responsible for damages to the furnace caused by their neglect. During the early years of the industry the drunkenness of the workmen had caused many accidents and losses.[49] There are

Stove Plate Cast at Durham Iron Works

several records on the ledgers of the ironworks where founders were debited for damages. For instance, at the Colebrookdale Iron Works, David Davies and John Chapman were charged £50 for "Damage done to ye ffurnace" through their neglect.[50] There is no record, however, to prove that this was paid and it might have been a sword held over their heads as an incentive to sobriety. Where the damages were for smaller sums there is evidence to show that they were deducted from wages.

The potters were relatively highly specialized workers who

worked at the furnaces only when small castings requiring a certain degree of skill were made. They were paid in 1786, four shillings per hundredweight, and allowed six pence a piece on hand ware, flasked ware, and on articles under twenty pounds in weight.[51] As most of the furnaces produced pig iron or the heavier type of castings, the potters were required infrequently, and only for short periods of time. For this reason, they often traveled from one furnace to another.[52] When the potters worked regularly, their wages averaged a little more than the earnings of the founders.

Among the other workers at the furnace were the gutterman, who had charge of the sand molds, two fillers or mixers, a wheeler, a stocktaker, and an ore-burner when it was necessary to burn the ore before smelting it if it contained sulphur. The fillers or mixers averaged a little more than £5 each per month. The pay of workers if free varied from £3 to £5 a month. The blacksmith received about £7 per month; the carpenter was paid about the same. The blacksmith's striker or helper received about £2 or a little more and board.[53]

The wages paid for blooming ore and drawing it into bar iron at the few bloomeries in operation in Pennsylvania during the eighteenth century, remained strikingly stable throughout the period. In 1728 and 1730, from four shillings and six pence to five shillings per hundredweight was paid for this work.[54] In 1786, Henry Drinker stated that the forgemen who bloomed ore and drew it into bars received five shillings for each hundredweight.[55] The output of the bloomery was about a ton and a half a week. The three forge workers therefore averaged a little less than £3 a week each.

At the refinery forges the rate of wages increased somewhat during the period, and there was a much greater variation in the different regions than was the case with the wages of furnace workers. In 1730, the finers received thirty shillings a ton when working single handed.[56] About the middle of the century, the same rate was paid at many of the ironworks.[57] In others such workers received as much as forty-five shillings when working single-handed.[58] There are some instances at this time of workers receiving seventeen shillings per ton and board, working single-handed, or eight shillings and six pence and board per ton when working double handed.[59] Toward the close of the period under discussion, finers received from forty shillings to forty-eight shillings a ton working single-handed.[60] To make a broad generalization, the rate of wages paid to the finers varied from an average of thirty shillings per ton at the beginning of the period to a maximum of forty-eight shillings a ton toward the close of the period. The total weekly wage of a finer working single-handed ran about £3 since two tons of bar iron could be made each week. Out of this amount he paid his helper. When working double-handed, the output was larger, running as high as three tons or more. At the same rate, each worker received a weekly wage of about £2 or a lit-

Tilt Hammer, Hay Creek Forge

tle more than £8 a month, but had no helper to pay. The wages of the chafery men who re-heated the anconies and hammered them into the finished bars did not vary materially from those of the finers.

Besides the finers and the chafery men, there were a few helpers at the forge for stocking charcoal and aiding the skilled workers. These must be classified as unskilled laborers. They received from two to three shillings a day. The forge carpenter was paid £7 a month, and the stocktaker, about fifty shillings per month "and found."[61]

The forgemen were held accountable for the amount of iron used in drawing out the bars. About 2,700 lbs. of pig iron was necessary to draw a ton of 2,240 lbs. of bar iron.[62] It should be noted that in the forging of bars, two processes were involved, first from the pig iron to the ancony, and second from the ancony to the finished bar. John Taylor, in a statement to the Lancaster County Court during a suit against one of his workmen, said that 2,200 lbs. of anconies should yield 2,000 lbs. of finished bars, and that it was customary for hammermen to pay for all iron used in excess of that amount.[63] Thus less than 300 lbs. was permitted as waste from the pig iron stage to that of the ancony, and about 200 lbs. from the ancony stage to that of the finished bar. For all waste in excess the forgeman was required to pay.

Many woodcutters were employed, for much wood had to be cut for making charcoal. The force of woodcutters was augmented during the winter season when the furnace was out of blast by many of the ironworkers. While often receiving a daily wage, woodcutters were usually paid by the cord. The rate varied but little from year to year. Down to the time of the Revolution, the rate remained about two shillings to two shillings and six pence a cord, although occasionally reaching two shillings and nine pence a cord.[64] In the years following the Revolution, the rate varied from two shillings and three pence to two shillings and six pence a cord, rarely going higher.[65] At the latter rate, a man chopping three cords of wood per day would earn seven shillings and six pence a day, or a weekly wage of two pounds, five shillings.[66]

By far the largest number of workmen required besides the woodcutters were the colliers, who "coaled" the wood cut by the woodcutters. As many as twelve were employed on plantations to keep a single furnace going.[67] The relationship here, again, between the ironmaster and the workmen varied. On some plantations, colliers received a daily wage. More frequently, agreements were drawn up and payment made by the bushel. For instance, in 1746, Rees Jones agreed to coal 200 cords of wood for John Taylor at eleven shillings and eight pence per 100 bushels to be paid for "half money, half goods as customary."[68] The agreements were usually made for 100 to 300 bushels.[69]

Teamsters were paid by the month or year. At Mount Joy (Valley) Forge they received £20 per year and board.[70] Written agreements were frequently made. In 1757, Conrad Wishon contracted with William Bird to "serve him as carter for 1 year at the rate of £19 per year, meat, drink, washing and lodging."[71] At the Warwick Iron Works, as at most of the iron-works, many of the teamsters who hauled the iron from the furnace to the various forges or to Philadelphia owned their own teams. They were paid by the ton for hauling the pig iron. While most of these workers were white, many of them were Negroes.[72]

The miners and their helpers were paid by the ton. While there were few shaft mines and although almost all mining was done in open pits, the miners had to be experienced, for their work required especially a knowledge of opening the veins of ore. They were well paid, receiving about three shillings and six pence per ton.[73] Each miner produced from one to two tons per day, giving him a daily wage from three shillings and six pence to seven shillings per day.

Among the other workers must be included the laborers, who did a variety of tasks at the ironworks and on the plantation. The wages varied from two to three shillings or more a day, or from forty shillings to sixty shillings a month.[74] The women workers were paid much less, and depended upon the work done, whether spinning, weaving, reaping, or other plantation work.[75]

It has been noted that wages varied to a great extent among the workers although the rate paid to the various classes changed but little over a long period of time. The lowest-paid laborer received about twelve shillings per week, while the skilled workers received a weekly wage of several pounds. In considering the rate of wages received by the workers, it should be noted that real wages on the iron plantations were greater than nominal wages, since wages were paid to the workers largely in goods. At the store on each plantation, goods were obtained at reasonable prices when compared with prices at Philadelphia and the boroughs, although the traveler, Hermelin, slyly remarks that those ironmasters who paid slightly higher wages than others charged their workers higher prices at the store for their supplies and merchandise. The ironmasters as a whole, however, dealt fairly with their employes, and the "truck system," which during the nineteenth century was so greatly abused as a means of reducing earnings and increasing profits, did not yet work to the disadvantage of the workers.

While the rate of wages of the different classes of workers

remained relatively stable during the period with a slight increase in some trades toward the later part of the century, the price of wheat at the plantation stores doubled between the years 1728 and 1756, from two shillings and sixpence to five shillings a bushel.[76] From about the middle to the end of the century, with the exception of the Revolutionary period, when prices became inflated,[77] the price remained about the same.

Country produce remained relatively cheap throughout the entire period. The workers themselves grew most of what they needed on small pieces of land allotted to them, or else secured their needs at the store. Eggs sold at four pence per dozen and chickens at six pence each. Salt pork, the principal meat eaten by the workers, sold at six pence a pound, dried beef at five pence, beef at four pence, and veal at two pence.[78] On the other hand, some articles were relatively expensive. Sugar sold at from eight pence to nine pence a pound. Tea and coffee were also expensive. In 1729, Bohea tea sold in Philadelphia at twenty-two shillings a pound.[79] On the whole, however, the common necessities of life were fairly reasonable in proportion to the wages paid. On special occasions, such as the reconstructing of a furnace stack, or when special exertion was required to complete a building, or when haste in the delivery of iron or other material was needed, allowances of rum were provided by the ironmaster.[80]

The necessities of life were obtained much cheaper at the plantation stores than in Philadelphia or in nearby villages. In 1730, wheat sold in the Philadelphia market as high as four shillings a bushel, while on the plantations the workers paid from two shillings and sixpence to three shillings a bushel.[81] The price of flour was also lower in about the same proportion when compared with Philadelphia prices. Nevertheless, the cost of living was much lower on the iron plantations than in the capital or in the villages and towns. It is true that fewer luxuries could be obtained at the stores of the ironmasters, but the workers were well provided with all the everyday necessities of life.

Prices of commodities in western Pennsylvania at the close of the century were only a little higher than they had been in southeastern Pennsylvania fifty years earlier, in spite of the fact that there could be found some who bewailed the high prices of the times.[82] In the Pittsburgh markets, flour sold at

$1.50 per cwt. buckwheat at $1.00 and oats at twenty cents a bushel. Beef sold at five cents a pound, pork at four cents, mutton at six cents, and venison at four cents. Potatoes were twenty-five cents a bushel, and butter thirteen cents a pound. Turkeys brought from forty to sixty cents each, ducks from twelve to seventeen cents, fowls from seven to ten cents, while eggs sold at seven to twelve cents a dozen.[83] With the exception of wheat and flour, the prices of commodities remained fairly stable throughout the century in the east, and after the settlement of the west, similar prices prevailed both in the east and the west.

During slack seasons, the workers could hunt game in the neighboring forests and could secure fish in the many streams. Some of the married workers, besides cultivating a small portion of land, kept a cow or two, some pigs, ducks, and chickens, and in this way reduced the cost of living. Fuel for domestic purposes could be obtained free or at a very small cost, since it was plentiful. House rent on the plantations was also comparatively low.

Few records remain of the wages paid to the workers in the slitting mills, plating forges, steel furnaces, and nail works of the period. From the fragmentary evidence that remains, the journeymen at the works in the towns and boroughs were paid far more than the laborers of the period, but not quite as much as the skilled workers at the furnaces and forges.[84] Although not living on the plantations, these workers also had the opportunity to cultivate small pieces of land in a period which was chiefly agricultural, and in this way could eke out a very comfortable living.

During the last decade or two of the century, the manufacture of cotton and wool cards which were made chiefly from iron wire, developed in Philadelphia. This marked the beginning of the employment of children, many of them not more than eight years of age, in the manufactures of Pennsylvania.[85] Many opposed the employment of children as well as women in the rising industries. Others defended child labor. Colonel David Humphries became lyrical over the use of children in industry:

> Teach little hands to ply mechanic toil
> Cause failing age o'er easy tasks to smile.[86]

Tench Coxe also replied to the critics who claimed that such work was demoralizing to the young. He asserted that vice was far worse in the seaport towns and in the courts than in the industrial towns. The wages paid to children, as might be expected, was quite low.

The scale of living of the workers in iron manufacture in Pennsylvania during the eighteenth century was proportionately far higher than that of ironworkers in European countries for the same period. The words of Tench Coxe, a writer who advocated the development of manufactures at the close of the century, can be applied especially to the status of the ironworkers during the entire period:

> Laboring people in the farming, manufacturing and mechanical trades can have constant employment and better wages than in the dearest countries of Europe . . . And though the wages of the industrious poor are very good, yet the necessaries of life are cheaper than in Europe and the articles used are more comfortable and pleasing.[87]

Few organizations of any kind, however, existed among those engaged in ironmaking. The variety of workers on each plantation, the isolation of the skilled workers on the various plantations, and the patriarchal relationship between the ironmaster and employes did not favor the growth of organizations or unions among the workers. From the earliest days of the Pennsylvania iron industry "clubs" were formed among the workers and small amounts paid to the ironmaster each month.[88] No records have been found of the activities of these, but they were evidently beneficial societies of some sort. The organization of industry in the towns into a loose guild system without the restrictions or advantages of European guild systems has been noted. By the close of the century, mechanical societies were being formed even in the West. A society of this type was established in Pittsburgh in 1788. Other organizations of like nature were started in different parts of the state.[89] The development of trade unionism among iron workers, however, was to come in the nineteenth century.

FOOTNOTES

1 From the days of Pastorious on, throughout the eighteenth century and into the nineteenth, writers pointed out the great scarcity of labor of all kinds. See F. D. Pastorius, *Province of Pennsylvania*, in Historical Society of Pennsylvania *Memoirs*, IV, Part 2, p. 95. Peter Kalm, *Travels into North America*, I, 389; "Achenwall's Observations on North America, 1767," *Pennsylvania Magazine of History and Biography*, XXVII, 14; Tench Coxe, *View of the United States*, p. 38. For difficulties of ironmasters in securing labor see Petition of Iron Masters, *Votes of Assembly*, III, 31; William Gwynn to George Ege, April 22, 1779, in C. B. Montgomery Collection of MSS.; *Fayette Gazette*, August 4, 1794; *The Union*, December, 1800.

2 Quotation in S. W. Pennypacker, *Hendrick Pannebecker*, p. 84.

3 Israel Acrelius, *History of New Sweden*, p. 165.

4 Thomas Penn to Rev. Mr. Barton, April 11, 1764, Penn Letter Book, VIII, 48-51.

5 *The Remarkable Case of Peter Hasenclever*, pp. 2, ff; *New York Journal or General Advertiser*, October 8, 1767.

6 Potts MSS. Colebrookdale, Warwick, Pottsgrove and other ledgers.

7 Lancaster County Historical Society, *Papers*, XVIII, 64.

8 *Pennsylvania Gazette*, April 16, 1740.

9 *Votes of the Assembly*, III, 432; *Pennsylvania Colonial Records*, IV, 469; Robert Proud, *History of Pennsylvania*, II, 221.

10 *Pennsylvania Statutes at Large*, V, c. 405; 29 George II, c. 35; *Pennsylvania Colonial Records*, IX, 17, 24.

11 *Ibid.*, X, 662.

12 *Ibid.*, X, 636.

13 Elizabeth Furnace MSS. Pig Iron Book, August 14, 1777; E. Wood to William Attley, Mary Ann Furnace, June 2, 1777, and June 23, 1777, Pennsylvania Miscellaneous Collection, February 1, 1777 to August 24, 1777 (Library of Congress); T. r. Ege, *History and Genealogy of the Ege Family*, p. 153.

14 Elizabeth Furnace MSS. Pig Iron Book, August 14, 1777.

15 T. P. Ege, *History and Genealogy of the Ege Family*, p. 153.

16 Furnace and Forge Ledgers. Compare with Israel Acrelius, *History of New Sweden*, p. 168.

17 The Federal Trades' Parade of 1788, in celebration of the ratification of the Constitution presents the best example of the development of this system as well as the progress of industrialism. In the parade were picturesque floats representing the various crafts. For an account of the Federal procession, see *Pennsylvania Packet and Daily Advertiser*, July 12, 1788.

18 The German craftsmen of Germantown, Lancaster, and other towns were famous for their handiwork throughout the entire period. B. Potts MSS. II Coventry, 1728 (1727-1734), p. 40; *Ibid.*, VII, Coventry, 1736 (1736-1745) p. 155.

19 *Ibid.* LXXI, Pottsgrove, 1772 (1772-1789), p. 35.

20 *Pennsylvania Gazette*, July 3, 1737; Potts MSS. XIII, 1743 (1743-1751), Colebrookdale, p. 168. The terms "indentured servants" and "redemptioners" were used loosely and often interchangably: "Narrowly, redemptioners meant those whose price of passage was paid on arrival, sometimes by friends, but more often by sale." C. A. Herrick, *White Servitude in Pennsylvania*, p. 4.

21 *American Weekly Mercury*, March 29, 1728; *Pennsylvania Gazette*, July 3, 1737, *Ibid.* April 16, 1761; Potts MSS. XXIII, Pine Grove Day Book, 1748 (1745-1757), p. 318.

22 *Pennsylvania Gazette*, September 15, 1773.

23 In 1728 a "servant man" was bought at Pottsgrove for £14 for four years' service; B. Potts MSS. II, Coventry, 1728; in 1755 at Pottsgrove £4-1-0 was paid for one years' service. Potts MSS. XL, Pottsgrove, 1755, p. 9; See C. A. Herrick, *White Servitude in Pennsylvania*, pp. 204. Such labor could easily be obtained at Philadelphia. The following is an example of an early advertisement: "Lately imported: A Parcel of likely Men and Women Servants and are to be sold by Samuel Ferguson at the Widow Fox's in Walnut Street, Philadelphia, on reasonable terms either for ready money, Country Produce, or Credit." *Pennsylvania Gazette*, 29th. 11th. Month, 1728.

24 While Negro slaves were employed at the ironworks and in many other occupations, there was continual opposition to their importation. From the beginning of the century, acts were passed placing restrictions on the traffic. In 1700, a maximum duty of twenty shillings was imposed on each Negro imported. *Pennsylvania Statutes at Large*, II, c. 85, section 2. Five years later this was doubled. *Ibid.* II, c. 164, section 5. In 1712 an almost prohibitive duty of £20 a head was placed by Great Britain because it interfered with the trade of the African Company. Board of Trade Papers: Proprieties, IX, Q 39, Q 42. Other laws imposing duties of varying amounts were passed from time to time: 1715—*Pennsylvania Statutes at Large*, III, c. 228, section 1; Board of Trade Papers: Proprieties, X (2), Q 159. 1720-1721—*Pennsylvania Statutes at Large*, III, 465; *Pennsylvania Colonial Records*, III, 38, 144, 171. 1725-26—*Pennsylvania Statutes at Large*, IV, c. 290, 291; *Pennsylvania Colonial Records*, III, 247-250. 1729—*Pennsylvania Statutes at Large*, IV, c. 304. 1761—*Ibid.* VI, c. 467; *Pennsylvania Colonial Records*, VIII, 575-576; *Votes of the Assem-*

FOOTNOTES—Continued

bly, 1761, pp. 25 ff. passim. 1773—*Pennsylvania Statutes at Large,* VII, c. 572; *Ibid.* VIII, c. 681; *Pennsylvania Colonial Records,* IX, 400-401, 443 ff. X, 72, 77. In 1780 when the act for the gradual abolition of slavery was enacted, the state of Pennsylvania forbade the importation of slaves. *Pennsylvania Statutes at Large,* X, c. 881.

25 *Votes of the Assembly,* III, 31.

26 *Pennsylvania Statutes at Large,* IV, c. 304.

27 Green Lane Forge Ledgers; Richard Backhouse MSS., Durham Ledgers; Potts MSS. LXX, Pine Forge, 1774 (1774-1781); New Pine Forge Ledgers, 1775-1778, pp. 87; ff. passim. Mount Joy Forge Ledgers, 1757-1765; Charming Forge Journal A, 1772-1775, Journal for Ledger A, No. 1, pp. 23 ff. *passim.;* When Thomas Potts leased Pine and Pottsgrove Forges for two years at a rental of £270 per annum, three Negroes were included. B. Potts MSS. XIV, Pottsylvania, 1758, p. 520.

28 *Ibid.* II, *Coventry,* 1728 (1727-1734) p. 152.

29 *Ibid.* LXXI, Pottsgrove, 1772 (1772-1789), p. 36.

30 *Ibid.* XXVIII, Pine Forge, 1748 (1748-1758), p. 59.

31 Israel Acrelius, *History of New Sweden,* p. 168.

32 Many entries such as the following may be found in the ledgers: "Gave Caesar for his Christmas Box, 1/—" B. Potts MSS. II, Coventry, 1728 (1727-1734), p. 237; *Ibid.* VII, Coventry, 1736 (1736-1745), p. 103. According to Acrelius, slaves were better treated in Pennsylvania than anywhere else in America. Israel Acrelius, *History of New Sweden,* p. 168.

33 *Pennsylvania Journal,* November 8, 1775.

34 Potts MSS. XIII, Colebrookdale, 1743 (1743-1751), pp. 76, 86, 100, *passim.*

35 Indian Accounts, C. B. Montgomery Collection of MSS.; Potts MSS. B II, Coventry, 1728 (1727-1734), pp. 11, 51, 76, 102; *Ibid.* IV, 1732 (1732-1740), pp. 42, 60, 64, 65, etc.

36 *Pennsylvania Packet,* October 7, 1778; *Philadelphia Federal Gazette,* September 1, 1796.

37 *Pennsylvania Statutes at Large,* VI, c. 486. Long before the ironworks were established, apprentices, as well as indentured servants, had been used in the Duke of York's settlement in Upland (Chester) and the surrounding districts. A law passed in 1676 provided that runaway apprentices and servants, upon being retaken, should serve double the time that they had been absent. Anyone aiding such runaways were liable to be fined £20 to be paid to the master and £5 to be paid to the court, while anyone harboring such persons were required to pay the master ten shillings for each day's concealment. *Duke of York's Book of Laws,* Sept. 22, 1676, p. 28. No general law, however, regulated apprenticeship during this period. When William Penn established his colony and laid out the plan of Philadelphia in 1682, the apprentice system was established immediately in the different handicrafts. In 1700 a law provided for the punishment of runaway servants and for a reward to be paid by the owner. *Laws of Pennsylvania,* 1682-1700, pp. 65-67. A provision of an act for the relief of the poor passed in January, 1705-6 authorized the overseers of the poor with the consent of two or more justices of the peace to bind out as apprentices, the children of poor parents for as many years as they saw fit. *Pennsylvania Statutes at Large,* II, c. 154. Other laws were passed which provided for the compensation of the owners of apprentices when such apprentices had run away to enlist in the militia, *Ibid.* II, c. 182, Section 11, while later laws required the written consent of guardians or masters before apprentices could be enrolled for military service. *Ibid.* V, c. 405, Section 2.

38 *Ibid.,* VII, c. 616.

39 *Philadelphia Federal Gazette,* November 29, 1793.

40 *Pennsylvania Packet and Daily Advertiser,* July 12, 1788.

41 New Pine Forge Ledgers.

42 Cole Book, New Pine Forge.

43 Potts MSS. XXVIII, Pine Forge, 1748 (1748-1758), p. 113; S. G. Hermelin, *Report About the Mines in the United States of America, 1783,* p. 61.

44 Colebrookdale Ledgers.

45 Potts MSS. IX, Mount Pleasant, 1748 (1739-1750), p. 59.

46 J. D. Schoepf, *Travels in the Confederation,* I, 198. Henry Drinker to Richard Blackledge, 10th month 4th, 1786, Drinker Letter Book, 1786-1790, p. 83.

47 *Ibid.,* p. 83.

48 S. G. Hermelin, *Report About the Mines in the United States of America, 1783,* p. 61.

49 *Pennsylvania Statutes at Large,* IV, c. 293.

50 Potts MSS., I, Pine Forge, 1729 (1720-1740), p. 47.

51 Henry Drinker to Richard Blackledge, 10th Month 4th, 1786. Drinker Letter Book, MSS., 1786-1790, p. 83.

52 C. B. Montgomery Collection of MSS.

53 Henry Drinker to Richard Blackledge, 10th Month 4th, 1786. Drinker Letter Book, 1786-1780. p. 83.

54 B. Potts MSS., II, Coventry, 1728 (1727-1734) pp. 54, 124.

55 Henry Drinker to Richard Blackledge, 10th Month 4th, 1786. Drinker Letter Book, 1786-1790, p. 83.

56 B. Potts MSS., II Coventry, 1728 (1727-1734) pp. 124 ff. *passim.*

FOOTNOTES—Concluded

57 *Ibid. XXIII,* Pine Forge, 1748 (1748-1759) p. 366. Compare with Israel Acrelius, *History of New Sweden,* p. 168.
58 Extracts and Memoranda of the Private Papers of John Taylor, Folio, Book A, Cope Collection, p. 138; Roxborough Iron Works, Account Book, pp. 5 ff. *passim.*
59 Potts MSS. XXIII, Pine Forge, 1748 (1748-1757), p. 369. *Ibid.* XL, Pottsgrove, 1755 (1755-1758), pp. 193 ff. *passim.* By "working double-handed" was meant that two men would take the place of each hammerman, refiner and chafery man.
60 Potts MSS., LXX, Pine Forge, 1774 (1774-1781), pp. 59 ff. *passim. Ibid.* LXXIII, Pine Forge, 1781 (1781-1782) pp. 14 ff. *passim.* Henry Drinker to Richard Blackledge, 10th Month 4th, 1768. Drinker Letter Book, 1786-1790, p. 83.
61 *Ibid.* p. 83.
62 Testimony of Edward Knight, Penn MSS. Document 27.
63 John Taylor *vs.* Caesar Andrew, May Term, 1754, in Extracts and Memoranda of the Private Papers of John Taylor, Folio, Book A, Cope Collection, p. 141.
64 Potts MSS. B. II Coventry, 1728 (1727-1734) pp. 52 ad. pass. *Ibid.,* Pine Forge, 1729 (1720-1740) p. 139; *Ibid.* Valley Forge, 1759 (1759-1763) p. 324. New Pine Forge Ledgers.
65 J. D. Schoepf, *Travels in the Confederation,* I, 198. Henry Drinker to Richard Blackledge, 10th Month 4th, 1786, Drinker Letter Book, 1786-1790, p.
66 There are only a few records of agreements made for cutting wood. One of the few found was made in 1732 between Samuel Nutt and a workman whereby the latter agreed to cut 160 cords of wood in return for one-third of the wood, "victuals, cloaths and furnish axes." Potts MSS. III, Coventry, 1730 (1730-1732) p. 329.
67 Henry Drinker to Richard Blackledge, 10th Month 4th, 1786, Drinker Letter Book, 1786-1790, p. 83.
68 Extracts and Memoranda of the Private Papers of John Taylor, Folio Book A, Cope Collection, p. 138.
69 *Ibid.,* p. 143.
70 Potts MSS., XLVII, Mount Joy Forge, 1759 (1759-1763) pp. 3 ff.
71 New Pine Forge Cole Book.
72 John Potts Receipt Book (Warwick) No. 90, 1762-1765.
73 Potts MSS., I, Pine Forge, 1729 (1720-1740), p. 232.
74 Furnace and Forge Ledgers.
75 An example of the variety of work performed by women is as follows: In 1751, Mary Fields received 6d a day for spinning, 1/— a day for making hay, and 2/3 a day for reaping. Potts MSS., Colebrookdale Ledger, No. 89, p. 14.
76 Potts MSS. I, Pine Forge, 1729 (1720-1740) pp. 95 ff. *passim; Ibid.* Pottsgrove XL, 1755 (1755-1758), pp. 76 ff. *passim.*
77 Extracts from the Rev. Dr. Muhlenberg's Journal, 1776-1777, Historical Society of Pennsylvania, *Collections,* p. 152; Henry Drinker to Richard Blackledge, 10th Month 4th, 1786, Drinker Letter Book, 1786-1790, p. 83.
78 New Pine Forge, Waste Book, April 10, 1760, April 22, 1760, June 16, 1760, etc. Potts MSS., XV Pottsgrove 1758 (1758-1769) p. 17; *Ibid.* II, Pine Forge, 1733 (1733-1750), p. 16. *Ibid.* Pottsgrove, 1755 (1755-1758), pp. 42 ff. *passim.* All the ledgers show about the same prices for these commodities throughout the period.
79 *Pennsylvania Gazette,* 3rd Month, 29th, 1729.
80 Potts MSS., II, Pine Forge, 1733 (1733-1750). C. B. Montgomery Collection of MSS.
81 Potts MSS., I, Pine Forge, 1729 (1720-1740) pp. 95 *passim. Ibid.,* II, Pine Forge, 1733 (1733-1750), p. 16; Lancaster County Historical Society, *Papers,* II, 240.
82 Tench Coxe, *View of the United States,* p. 485.
83 *Pittsburgh Gazette,* January 2, 1801.
84 B. Potts MSS., III, Coventry, 1730 (1730-1732) pp. 132-178 *passim.*
85 *Universal Asylum and Columbian Magazine* (1791) II, 135-136.
86 David Humphreys, *Poem on Industry.*
87 Tench Coxe, *View of the United States,* p. 95.
88 Potts MSS., I, Pine Forge, 1729 (1720-1740) pp. 25, 51, 59 ff. *passim.* The ledgers of the ironworks of the Schuylkill Valley contain many references to "clubs."
89 *Pittsburgh Gazette,* March 22, 1788, February 4, 1797; C. W. Dahlinger, *Pittsburgh—A Sketch of Early Social Life,* pp. 18, 19.

CHAPTER VII

THE IRONMASTERS

THE ironmasters formed almost as diverse a group as the workers. The first men to establish ironworks were British and during the early years of the period the larger number were by far of the same origin, although as time went on the ironmasters represented several different countries. Thomas Rutter, Samuel Nutt, William Branson, Thomas Yorke, John Ross, John Taylor and others had set sail from the mother country to seek their fortunes in the new world. Among those of Welsh origin, James Morgan, Thomas Potts, David Jenkins, and James Old were outstanding. The German ironmasters included John Lesher, Henry William Stiegel, Gerrard Etter, John Probst, and Peter Shoenberger. Valentine Eckert came from Hanover, and George Anshutz from Alsace. Of the Scotchmen who engaged in iron manufacture, Sir William Keith, Samuel McCall, George McCall, and James Wilson were prominent, the latter, however, because of his public services. Robert Grace, Robert Coleman, and George Taylor were of Irish descent. Peter Marmie was a Frenchman and William Dewees was of French ancestry. In many cases, with the advance of time the sons of these men as typified by the Potts, Rutter, and other families, learned the business of ironmaking and continued the various enterprises established by their forebears. By the beginning of the nineteenth century most, but not all, of the important industrial leaders were native-born Pennsylvanians.

It is not possible to make a generalization of a single class from which the ironmasters were drawn, because they came from different social groups. Robert Grace was one of the very few who could lay claim to being a descendant of British aristocracy. He belonged to the family of the Barons of Courtstown and Lords of Grace's county. Members of the family were among the earliest Anglo-Norman settlers in Ireland. As early as the twelfth century the family received extensive grants of lands in Kilkenny county and prospered there for almost five centuries. Owing to their loyalty to the cause of the Stuarts, the estates of the family were confiscated during the Glorious Revolution.[1] William Keith, surveyor-general of customs in the colonies during the reign of Queen Anne, later became deputy governor of Pennsylvania, and succeeded to the baronetcy in 1721.[2] He was much interested in iron manufacture as well as in the encouragement of other branches of industry, but his career as ironmaster was short-lived. Only a small group, however, associated with iron manufacture in the colonies came from the ranks of the British upper classes.

Many merchants were attracted to iron manufacture, a few becoming ironmasters, but most of them only furnishing capital toward the establishing of ironworks. In fact it seems quite evident that much of the capital required in the rising iron industry was created in commercial and business enterprises.[3] William Branson, Samuel McCall, John Potts, Joseph Turner, Robert Ellis, George Fitzwater, William Allen, Clement Plumsted, Jeremiah Langhorne, John Hopkins, Charles Read, Anthony Morris and others were merchants who invested in ironworks or built them. By the close of the century a large number of Philadelphia merchants, such as Daniel Drinker and Thomas Clifford, Jr., had investments, more or less extensive in iron manufactures. When the industry took root in the western part of the state, merchants were among the foremost promoters. William Turnbull, Peter Marmie, and Robert Holker, prominent merchants of Pittsburgh, became partners in the Alliance Iron Works on Jacob's Creek.[4] These are but a few examples of many merchants who invested money in iron manufacture.

Of those who became prominent ironmasters, many had risen from the lower ranks through constant toil and unabating thrift. The pioneer ironmaster, Thomas Rutter, had been a blacksmith,[5] Robert Coleman, a clerk at Quitapahilla Forge,[6] Samuel Savage, Jr., a forgeman at Coventry Iron Works. David Jenkins had been employed as a clerk by Lardner, Flower and Hockley at Windsor Forges. In 1773, he purchased a half interest in the company for £2,500. Two years later, he bought the remaining half interest.[7] Such men mastered all phases of ironmaking and usually became very successful iron-masters.

The outstanding example of one who rose from the ranks was that of Robert Coleman. As a clerk at Quitapahilla Forge, he saved a sufficient sum of money to buy an interest in Salford Forge in 1773.[8] A few years later he leased Elizabeth Furnace which within a short time he owned outright. Writ-ing of his rise, he stated:

> In the year 1776, possessed of but a small capital and recently married, I took a lease for the Elizabeth Furnace estate for the term of seven years, not anticipating at that time that before the expiration of the lease I should have it in my power to become owner in fee simple of the whole or a greater part of the estate.[9]

Before the century closed he purchased interests in Cornwall Furnace,[10] Mount Hope Furnace, and Hopewell Forge.[11] He also built Colebrook Furnace.[12] When he died on March 8, 1825, he owned most these ironworks.[13]

A number of ironmasters were of "yeoman" origin. Farmers who were fortunate enough to find ores on their property, often with the aid of a number of partners, built ironworks. The following record, taken from a deed book illustrates the point and is but one of many similar entries that can be found in the records of the eighteenth century: In 1767, "Jacob Shöffer of Manatawny, yeoman," for a consideration of £500 sold an undivided fourth part of a tract of 175 acres of land situated on Moselem Creek, Richmond township, together with "one fourth part of all forges, mills, etc., erected thereon" to Christian Lower of Tulpehocken, a blacksmith. Such "yeomen" frequently had difficulty in securing the necessary cash to ful-fill their obligations to operation of the ironworks.

The close connection between ironmaking and agriculture, owing largely to the needs for fuel was a relationship that continued well into the nineteenth century. It was natural that well-to-do plantation owners should use a part of their wealth to engage in an industry so closely allied to agriculture. Thus capital that accumulated in farming found its way into industrial enterprises. For example, in his will, Cyrus Jacobs, who owned Pool and Spring Grove Forges in Lancaster county, styled himself as "iron master and farmer."[14] In many eighteenth century documents a number of instances of this sort may be found.

Many of the outstanding civic and political leaders, while not actively engaging in the iron industry, owned interests, often large, in various works. John Dickinson had an interest in Elizabeth Furnace, the plant of the unfortunate Henry William Stiegel.[15] Joseph Galloway, the first individual owner of the Durham Iron Works, was a member of the First Continental Congress and presented a compromise plan of dealing with the situation which confronted the colonies.[16] His adherence to the mother country cost him his life interest in his ironworks and his vast estate.[17] Michael Hillegas, appointed the first Treasurer of the United States by the Continental Congress, held a large interest in Martic Forge, Lancaster county.[18] James Wilson, in partnership with his brother-in-law, Mark Bird, owned ironworks at the falls of the Delaware River, including a forge and slitting mill.[19] Before his death, he also came into possession of Hopewell Furnace.[20] Although Benjamin Franklin was very much interested in the iron industry, especially perhaps because his friend Robert Grace owned Warwick Furnace, he did not invest any of his money in this manner. When friends sought his advice in this respect, the wise old philosopher counseled against such investments stating that as a result of such speculation they might "only meet with disappointment."[21]

Among those who held public office in Pennsylvania during the eighteenth century are the names of many ironmasters. Samuel Nutt was a member of the Provincial Assembly from Chester county from 1723 to 1727.[22] Thomas Rutter, Jr., represented Philadelphia county from 1727 to 1730.[23] Other representatives who served in the Assembly included Thomas Yorke (Berks county),[24] Samuel Potts (Philadelphia county),[25] Matthias Slough (Lancaster county),[26] Curtis Grubb (Lancaster county), Thomas Potts (Philadelphia county),[27] and Valentine Eckert (Berks county).[28] Others who were elected to the Assembly at various times during the Revolution were Mark Bird, Christian Lower, George Ege, and John Patton.[29]

Many of the judges were drawn from the ranks of the ironmasters. Thomas Yorke,[30] John Potts,[31] Samuel Potts,[32] Valentine Eckert,[33] James Old[34] and Isaac Meason,[35] George Stevenson and Richard Backhouse[36] sat on the bench. Appointed jus-

Six Plate Draft Stove

tices of the peace, these men were also judges of the county courts. The brilliant and distinguished James Wilson, already mentioned, who became an associate justice of the Federal

Supreme Court, being appointed by Washington, must also be included among those having extensive interests in various ironworks. For several terms Wilson was a member of the Continental Congress. He was a signer of the Declaration of Independence and an important member of the Constitutional Convention of 1787. William Allen was chief justice of the Supreme Court of Pennsylvania. Among the members of the Governor's Council was James Logan. Lynford Lardner for many years was also a member of the council.[37] Francis Rawle was appointed but did not accept.[38] John Taylor was the sheriff of Chester county for many years,[39] while George McCall served as a member of the Common Council of Philadelphia.[40]

Especially during the Revolution and following it, the ironmasters occupied a very important place in public life. The part played by Joseph Galloway, the first individual owner of the Durham Iron Works as a member of the First Continental Congress has already been mentioned. George Taylor, who leased the Durham Works, James Smith of Codorus Iron Works, George Ross of Mary Ann Furnace, as well as James Wilson were among those who signed the Declaration of Independence. Valentine Eckert, the owner of Windsor Furnace and later Sally Ann Furnace, David Jenkins of Windsor Forges, Mark Bird of Birdsboro Iron Works and James Smith of Codorus Iron Works, were members of the Provincial Conference of 1776.[41] The ironmasters were well represented on the important committees and at the meetings held in Philadelphia prior to, and during the struggle with the mother country.[42]

The membership of the Pennsylvania Constitutional Convention of 1776 included such industrialists as Valentine Eckert, John Lesher, Thomas Potts, and George Ross.[43] The industrialists were likewise represented in the Constitutional Convention of 1790.[44] A number of ironmasters like John Nicholson, who became comptroller-general, held important offices in the state from time to time. Among those who sat in the Pennsylvania legislature were Daniel Udree, Christian Lower, Thomas Potts, Thomas Bull, Matthias Slough and others.[45] In 1813, Daniel Udree, the proprietor of Oley Iron Works in Berks county, took his seat in the United States Congress.[46]

A glance backward to the period before the Revolution shows that ironmasters and their sons held commissions in the provincial service.[47] Even during the French and Indian War when the public attitude, especially on the part of the Quakers, was largely one of indifference in spite of the fact that the frontiers were continually in danger, a few ironmasters marched at the head of their regiments against the enemy.

In the period following the French and Indian War, as difficulties grew between the colonies and the mother country, many of the ironmasters enlisted in the ranks of the most prominent radicals. Several like David Potts took a leading part in the non-importation agreements.[48] The ironmasters had grievances against Great Britain especially on account of the prohibitive clauses of the Act of 1750 and because iron became an enumerative product in 1764,[49] although both these laws were almost entirely disregarded. Also, as secondary iron manufactures developed, the Americans found themselves competing with British manufacturers. These grievances must be added to the other fundamental causes of the Revolution. The leadership and aggressiveness of the owners of the ironworks, together with their services, must be taken into account in a consideration of the struggle for independence.

The work of many of the ironmasters in the political phases of the Revolution has already been pointed out. Almost without exception, as a group, the ironmasters stood for the American cause from the very beginning of the conflict and were among the most active in the struggle against the mother country. Joseph Galloway and John Potts were exceptions.[50] Not only were the ironworks kept running at full speed turning out cannon, shot and munitions for the armies, but many ironmasters as commissioned officers saw active service in the battles of the Revolutionary War.

Among those who took part in the military activities of the war were Lieutenant Daniel Udree of the Oley Iron Works, prominent in the hard-fought Battle of the Brandywine.[51] He also served in the War of 1812 as a major-general.[52] William Thompson of Mary Ann Furnace became "Colonel of the Battalion of Rifle Men Raised in the Province of Pennsylvania"[53] in June, 1775, and later took part in the Canadian campaign as brigadier-general. At the dreadful disaster of Three Rivers, he was taken prisoner, but was later exchanged. He died in Carlisle in 1781 before the war had been brought to a successful conclusion. Another meritorious officer from the

General Daniel Udree, Ironmaster, Oley Furnace

138

same region as Thompson was Philip Benner, who became an important ironmaster after hostilities had ceased.

Many other ironmasters were officers in the American army during the Revolution. Included among the colonels were George Ross, Matthias Slough, Peter Grubb, Curtis Grubb, Thomas Potts, John Bull, Thomas Hockley, James Smith, Mark Bird, Christian Lower, and George Taylor. Among others who served were John Patton, Joseph Thornburg, John Lesher, Robert Coleman, James Chambers, and Benjamin Chambers. William Dewees, who was associated with Isaac Potts at Valley Forge at the time it was burned by the British, was also an officer. Among the commissioners for purchasing provisions for the army were the ironmasters John Lesher, Valentine Eckert, and Christian Lower.[54] Almost every ironmaster served in some active capacity during the struggle that led to independence.

From time to time during the Revolution even those who adhered to the American cause suffered from the manner in which supplies were taken for the army. The ironmasters were not exempt from these raids as is evidenced in the following extract of a letter written by Jacob Lesher to the president of the Supreme Executive Council:

> I conceive it to be my duty to acquaint you that I am no more master of any individual thing I possess! for besides the damages I have heretofore sustained by a number of troops and Continental wagons in taking from me 8 tons of hay, destroyed apples sufficient for 10 hhds. of cider, eating up my pasture, burning my fences, etc. and two beeves, I was obliged to buy at 1 sh. per lb. to answer their immediate want of provisions, and at several other times, since, I have supplied detachments from the army with provisions. There has been lately taken from me 14 head of cattle and 4 swine. The cattle at a very low estimate, to my infinite damage, as they were all the beef I had for my workmen for carrying on my iron works.
>
> I had rather delivered the beef and reserved the hides, tallow, etc., but no argument would

prevail! all must be delivered to a number of armed men at the point of the bayonet. As my family, which I am necessitated to maintain consists of nearly thirty persons, not reckoning colliers, wood cutters, and other day laborers! My provision and forage being taken from me, my forge must stand idle! My furnace (which I am about carrying on) must of consequence be dropped! which will be a loss to the public as well as myself as there is so great a call for iron at present for public use, and some forges and furnaces must of necessity fail for want of wood and ore. The case in this neighborhood is truly alarming when the strongest exertion of economy and frugality ought to be practiced by all ranks of men! . . .[55]

The complaint was a general one, for during the exigencies of war, especially during the first years of conflict, supplies and provisions were taken by the troops regardless of ownership.

While many ironmasters prospered during the eighteenth century, a tragic phase of the Pennsylvania iron industry was the large number of failures. The problem of securing capital in sufficient quantities continually faced those who engaged in ironmaking. Besides this, the difficulties in obtaining cash payments for the iron sold, the problem of securing skilled labor, the locating of furnaces near poor ores, or where the supply ran out, the high cost of transportation from the plantations, a few of which were almost inaccessible before roads could be built, and mismanagement in the conduct of the business, were causes, one or more of which led to the downfall of many ironmasters.

About the middle of the eighteenth century, Acrelius, in writing of Crosby's or Crum Creek Forge, stated that it had "ruined Crosby's family."[56] James Smith, who bought Codorus Iron Works on Codorus Creek near the Susquehanna River, in 1771, lost £5000 in his venture. He attributed his misfortune to his two managers, one of whom he said was a knave, and the other a fool.[57] Turnbull, Marmie and Company, the owners of the Alliance Iron Works on Jacob's Creek, failed not many years after they began iron manufacture.[58] Their difficulties were largely due to the fact that they could not collect money owing to them. They had large claims "against the government of France for advances made to the Royal Marine."[59] Among the long list of ironmasters who became bankrupt must be included Matthias Slough, James Old, Frederick Delaplank, John Truckenmiller, Henry William Stiegel, and Mark Bird.

The most pathetic of all the failures among the ironmasters was that of Henry William Stiegel, who came to America from Rotterdam in the ship *Nancy* with his widowed mother and young brother. "Heinrich Wil Stiegel" as he designated himself on the register at McCall's wharf in Philadelphia, rose rapidly until he became an ironmaster and the owner of glass works. A few years after coming to the New World, he married Elizabeth, the daughter of Jacob Huber, an early Lancaster county industrialist. In 1756, with a number of partners, Charles Stedman, Alexander Stedman, and John Barr, he took over the Huber furnace.[60] A new furnace was erected soon afterwards and called Elizabeth Furnace. At this works a variety of cast ironwares were made from kettles, pots and pans to "cast funnels of any height for refining sugars, weights of all sizes, grate bars and other castings for sugar works in the West Indies."[61] Stiegel bought Charming Forge,[62] some miles distant from Elizabeth Furnace and later the Stedmans became part owners also.[63] Beginning in 1763, Stiegel began erecting glass houses at Manheim.[64] He laid out the village of Manheim and secured skilled labor from Germany for his glass works. Famous glassware was produced, which today because of its beauty is keenly sought by collectors.

During the height of his prosperity, this ironmaster, who loved display, lived in a very ostentatious manner. His mansion, richly finished, was located at Manheim. The platform surmounting the gabled roof was modeled like a grandstand for his workman orchestras to welcome him when he dashed home from Philadelphia, Charming Forge or Elizabeth Furnace in his coach and four. He erected two towers where cannon were mounted to salute his arrivals and departures by the burning of powder. All these point to a man who loved splendor and show.[65]

Like many ambitious speculators of his period—in lands,

agriculture or industry—this ironmaster, town builder and glassmaker planned too rapidly. For before 1768, Stiegel and his partners, the Stedmans, were deeply in debt.[66] Stiegel hoped to obtain £8,000 for his interests in Elizabeth Furnace and Charming Forge, but the properties could not be sold.[67] He then mortgaged all his holdings, including his scattered lands, to Daniel Benezet of Philadelphia, for £3,000.[68] It is evident that Stiegel was now planning to concentrate all his energies on his new glass plants.[69]

Stiegel's financial affairs became more and more involved. His entire Manheim estate was mortgaged in 1770 to Isaac Cox for £2,500.[70] The next year he petitioned the Assembly for aid, stating that he had been put to a great expense in building his glass works and in procuring skilled workmen from Europe.[71] A year later he received £150 from the Assembly.[72] In 1773, a lottery was held for his benefit. Finally, however, his creditors pressed him for payment, his works were advertised by the sheriff and he was placed in jail for debt.[73] Once again he petitioned the Assembly, but this time for freedom.[74] This was granted by a special act of the Assembly.[75] During his remaining years he taught school and for a short time became a clerk for George Ege at Charming Forge. He died in poverty in 1785.

The memory of this erratic ironmaster still lives in the annual anniversary of the "Feast of Roses." This observance grew out of the provisions of a gift of land to the German Lutheran Congregation at Manheim, the deed providing that there should be paid to him "his Heirs and Assigns at the Town of Manheim in the month of June yearly forever hereafter the Rent of One Red Rose if the same shall be lawfully demanded.[76] The deed came to light in 1891 and the "Feast of Roses" which has been observed every year since brings back the memory of a man who had reached the heights of success and then failed. His downfall may be accounted for, in part at least, by too rapid an expansion on a small capital. His books and ledgers, chiefly kept by himself, were never in good order, and he could never tell where his business stood. A great deal of money was also owing to him which must be considered in accounting for his failure.[77]

Another outstanding and pathetic failure was that of Mark Bird. After his father's death in 1761, he took charge of the estate[78] and finally came to own all the Birdsboro forges. He expanded his holdings there. He also built Hopewell Furnace. During the drab years of depression which followed the Revolution, he became financially involved. His brother-in-law, the prominent James Wilson, became his partner. Their vast interests, including Birdsboro, Spring Forge and many tracts of land were mortgaged in 1784 to John Nixon for 200,000 Spanish milled silver dollars.[79] The properties were sold by the sheriff in 1780,[80] although James Wilson subsequently owned a

George Taylor

part of them.[81] Mark Bird left Pennsylvania to live the concluding years of his life in North Carolina.[82]

In contrast to the many tragic failures were the industrialists who were successful. The Potts family might be cited as an exceedingly prosperous group of ironmasters.[83] The valuation of the personal estate alone of John Potts, who died in 1768, amounted to £7586-14-4, a large sum for his day. Besides this, and his interests in ironworks in the Schuylkill Valley, he was a partner in a Virginia iron plantation, and also owned large tracts of land in Path Valley, Cumberland County.[84] The meteoric rise of many who engaged in iron manufacture bears witness to the fact that if the business was rightly conducted, with favorable conditions, success was assured.

Matrimonial alliances connected many families of ironmasters. The outstanding example of such inter-marriages is afforded by the history of many of the ironmasters of the Schuylkill Valley. The unions between the Potts, Rutters, and Savage families tended to keep a large part of the industrial wealth in the hands of a few families in that region. Among the ironworks owned by the united Potts and Rutter families, for instance, were Colebrookdale Furnace and Forge, Mount Pleasant Furnace and Forge, Warwick Furnace, Coventry Iron Work, Rutter's Forge, Spring Forge, Pool Forge, Pine Forge, Little Pine Forge, McCall's Forge and others.[85]

Another example, in the Lancaster county region, of "iron marriages" was between the Stiegel, Old, and Ege families. James Old had two noted ironmasters as sons-in-law, Cyrus Jacobs who built Spring Grove Forge, and Robert Coleman. William, a son of James Old, married Elizabeth, the daughter of Henry William Stiegel, and their son, Joseph, married Rebecca, the only daughter of George Ege.[86] There were other alliances between these families. The tendency among many families of ironmasters to intermarry resulted in bringing wealth, with increased opportunities for industrial development.

Most of the ironmasters were men of education and culture, although many of the first generation, being of lowly origin, had not the opportunities which they could afford their sons. A tutor or school master provided instruction and the sons of many ironmasters were sent to Europe to finish their education.

While little was known about the chemical properties of iron and steel, and although the study of metallurgical chemistry was unknown, several ironmasters obtained a training in the best schools of Europe in mineralogy and "the fluxing of metals." Robert Grace, the genial ironmaster at Warwick, had received such training and used his knowledge to advantage at his mines and ironworks.[87] Many members of the Potts family of ironmasters, likewise, were skilled in the knowledge of metals and mining. Colonel Thomas Potts of Pottstown and Coventry was noted in this respect.[88] There was little opportunity, however, at this early period for the development of theories and laboratory experimentation, although discussions of various kinds were held at the American Philosophical Society in Philadelphia. Dr. Joseph Priestley's doctrine of phlogiston was attacked by Dr. James Woodhouse, professor of chemistry at the University of Pennsylvania, during a discussion of the composition of "finery cinder" produced at the forges.[89] The study of metallurgy, however, was but in its infancy.

In a new country with a scattered population, it was perhaps natural that ironmasters did not form associations as was being done in England at this time. In periods of stress, however, calls were sent out for meetings where problems were discussed and policies formulated. A few years before the Revolution when there was great difficulty in obtaining money to carry on iron manufacture, the New Jersey ironmasters sent out a call to the owners of furnaces and forges in that colony and the neighboring colonies.[90] A meeting held at Moristown sent out a second call to the ironmasters of New Jersey, New York, and Pennsylvania to meet on the 29th day of December, 1773, for the purpose of "forming some Regulations which seems absolutely necessary in carrying on Iron Works in this part of America."[91] In the various local regions, likewise, the ironmasters occasionally called meetings to discuss their affairs. Such notices as the following were occasionally placed in the newspapers: "The different Iron Masters are hereby requested to meet at the House of Mr. Michael Wood, Innkeeper in the Borough of Reading, on Wednesday the 7th day of January next at 11 o'clock in the forenoon to consult on business of importance. December 30, 1794."[92] There are indications to show that the question before this meeting was the tariff. Little progress, however, was made throughout the century in organizations of any kind either among the groups of ironmasters or within the ranks of the workers.

FOOTNOTES

1 J. B. Burke, *Peerage and Baronetcy*, pp. 609-611.

2 William Keith, *Just and Plain Vindication;* pp. 7-8; Charles P. Keith, "Sir William Keith," *Pennsylvania Magazine of History*, XII, 1-33.

3 The vast amount of correspondence and other manuscript material at the Historical Society of Pennsylvania testifies to these activities of many Philadelphia merchants.

4 *Pittsburgh Gazette*, April 12, 1788, November 22, 1788, November 2, 1793.

5 Jonathan Dickinson to James Logan, 1717, Dickinson Letter Book, 1713-1721.

6 *Alden's Appeal Record*, Coleman *vs.* Brooke, p. 152.

7 Lancaster County Historical Society *Papers*, XVIII, 61.

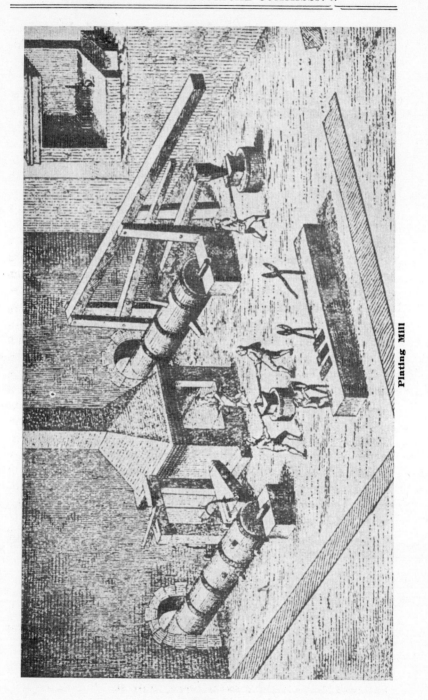

Plating Mill

8 While at Salford Forge, Robert Coleman received the order from the Committee of Safety to forge the chain designed to span the Delaware for defense against the British fleet. *Pennsylvania Archives*, First Series, V, 39.

9 Robert Coleman, History of the Elizabeth Estate (MSS.), Quoted in J. M. Swank, *Introduction to the History of Ironmaking and Coal Mining in Pennsylvania*, p. 19.

10 Lancaster County Deed Book, II, pp. 35 ff.

11 *Ibid.*, c. III, 514 ff; Muniments of Title No. 16.

12 Lebanon County Historical Society *Papers*, III, No. 1, p. 7.

18 Lancaster County Wills: Robert Coleman, Will dated March 8, 1822, proved September 3, 1825.

14 *Ibid.*, Will of Cyrus Jacobs, May 5, 1830.

15 Henry W. Stiegel to John Dickinson, June 24, 1771, Logan Papers, XXXVIII, Ducuments 87.

16 *Journals of the Continental Congress*, I, 43-48. The twelve original owners formed a trust in 1727 to continue for fifty-one years. Philadelphia County Deed Book, G, III, 240-245. A Deed of Partition was executed December 14, 1773, Galloway receiving the Durham Furnace tract.

17 *Pennsylvania Colonial Records*, XII, 104.

18 Lancaster County Deed Book, G.

19 James Wilson Papers, VII, 21; *Pennsylvania Packet and Daily Advertiser*, July 8, 1788.

20 *Reading Weekly Advertiser*, May 27, 1797.

21 Benjamin Franklin to M. Bernard, September 12, 1784, Miscellaneous Papers and Letters of Dr. Franklin: Miscellaneous, 1783-1788, LV, Part 2, Document 76.

22 *Votes of the Assembly*, II, 374-494 *passim*. III, 3-35 *passim*.

23 *Ibid.*, III, 36-124, *passim*.

24 *Ibid.*, IV, 625-856, *passim*.

25 *Ibid.*, VI, 1-261, *passim*.

26 *Ibid.*, VI, 483-766, *passim*.

27 *Ibid.*, VI, 622-766, *passim; Journal of the Assembly*, I, 526-698, *passim*.

28 *Ibid.*, I, 97-158, *passim;* 390-525, *passim*.

29 *Ibid.*, I, 232 ff. *passim*.

30 *Pennsylvania Colonial Records*, VIII, 575. *Pennsylvania Archives*, Third Series, IX, 252.

31 *Ibid.*, IX, 252-253 ; *Pennsylvania Colonial Records*, VIII, 575.

32 *Pennsylvania Archives*, First Series, VI, 595.

33 *Pennsylvania Magazine of History and Biography*, III, 323.

34 *Pennsylvania Colonial Records*, XV, 52.

35 Fayette County Deed Book, B, p. 149.

36 *Pennsylvania Gazette*, June 16, 1768, *Colonial Records*, XV, 52.

37 *Pennsylvania Archives*, Second Series, IX, 624.

38 *Pennsylvania Colonial Records*, III, 232. Francis Rawle was one of the builders of Pool Forge in the Manatawny region. Pool Forge Account Book, 1725-1727.

89 Extracts and Memoranda of the Private Papers of John Taylor, Folio A, Cope Collection, p. 140.

40 *Pennsylvania Archives*, I, 203-204.

41 Proceedings of the Provincial Conference, June 18-June 25, 1776, MSS.

42 *Journals of the House of Representatives*, I, 4 ff.

43 *Pennsylvania Magazine of History and Biography*, III, 323-324 ; IV, 89, 226-227, 230-231.

44 *Pennsylvania Archives*, First Series, XII, 27.

45 *Minutes of the Assembly*, 1781-1790.

46 *Pennsylvania Archives*, Fourth Series, IV, 816-817.

47 The names of many of these may be found in the rolls of officers of regiments in the various counties. See *Pennsylvania Archives*, Second Series, II, 502 ff. etc.

48 Mrs. T. P. James, Manuscript Collections.

49 23 George II, c. 29 ; 4 George III, c. 15.

50 Mrs. T. P. James. *Memorial of Thomas Potts, Jr.*, p. 159.

51 *Pennsylvania Archives*, Fifth Series, V, 26, 27, 125, 133, 175, 189.

52 *Ibid.*, Sixth Series, VII, 4.

FOOTNOTES—Concluded

[53] Commission of William Thompson, Carlisle Public Library.

[54] *Pennsylvania Archives,* First Series, VI, 327.

[55] John Lesher to President, Supreme Executive Council, January 9, 1778, *Ibid.,* First Series, VI, 170.

[56] Israel Acrelius, *History of New Sweden,* p. 169.

[57] W. C. Carter and A. J. Glossbrenner, *History of York County,* Appendix, p. 10.

[58] *Pittsburgh Gazette,* November 2, 1793.

[59] Fayette County Deed Book, C, III, 1319.

[60] Elizabeth Furnace Ledgers; Philadelphia County Deed Book, I, IV, 111-113.

[61] *Pennsylvania Gazette,* March 23, 1769; May 4, 1769.

[62] Philadelphia County Deed Book, I, IV, 97.

[63] Charming Forge Account Book, 1763. Entry February 5, 1763.

[64] Ledger A No. 1 for Manheim Glass House.

[65] F. W. Hunter, *Stiegel Glass,* Chapters III-VI.

[66] *Pittsburgh Gazette,* March 24, 1768.

[67] Henry William Stiegel, Proposal to Sell Elizabeth Furnace and Charming Forge, Logan Papers, XXXVIII, 72.

[68] Lancaster County Mortgage Book, N, p. 23.

[69] His advertisements seem to show that he was planning to give up the manufacture of iron. *Pennsylvania Gazette,* March 23, 1769, May 4, 1769; *Pennsylvania Journal and Weekly Advertiser,* July 5, 1769, August 9, 23, 1769, Day Book. Manheim Glass Works MSS.

[70] Lancaster County Mortgage Book, N, pp. 434 ff; O, pp. 219 ff.

[71] *Votes of the Assembly,* VI, 147.

[72] *Ibid.,* VI, 400.

[73] *Pennsylvania Gazette,* December 28, 1774.

[74] *Votes of the Assembly,* VI, 554.

[75] *Ibid.,* VI, 559, 560. *Pennsylvania Statutes at Large,* VIII, c. 702; Warrant in Penn-Physick Papers, XV, 69.

[76] F. W. Hunter, *Stiegel Glass,* p. 117.

[77] The Elizabeth Furnace Ledgers and Manheim Glass House Account Books are at the Historical Society of Pennsylvania.

[78] Anna Krick MSS.; *Alden's Appeal Record,* Coleman *vs.* Brooke, p. 441.

[79] James Wilson Papers, VII, 21. *Ibid.* III, 24.

[80] *Pennsylvania Packet,* March 26, 1788.

[81] *Reading Weekly Advertiser,* May 27, 1797.

[82] His petition in bankruptcy was refused by the North Carolina legislature. *Colonial and State Records of North Carolina,* XXI, 760, 779, 909, 914, 933, 952.

[83] Mrs. T. P. James, *Memorial of Thomas Potts, Jr.* p. 115.

[84] Philadelphia County Wills: John Potts, Book O, pp. 245 ff.

[05] Mrs. T. P. James. *Memorial of Thomas Potts, Jr.,* p. 42.

[86] T. P. Ege, *History and Genealogy of the Ege Family.* See also *Stiegel Genealogy,* F. W. Hunter, *Stiegel Glass,* Appendix.

[87] P. Colinson to John Bartram, August 17, 1737, Bartram Papers, II, Document 38.

[88] Mrs. Thomas Potts James, Manuscript Collections.

[89] American Philosophical Society Transactions, IV, 466-469.

[90] *New York Gazette and Weekly Mercury,* August 23, 1773.

[91] *Ibid.,* December 20, 1773.

[92] *Readinger Zeitung,* December 31, 1794.

CHAPTER VIII

Relations with England

ABOUT the time that the foundations of the first bloomery forge were laid in the Manatawny region of the Schuylkill Valley, attempts were made by various groups in England to control the colonial iron industry, which by this time was beginning to take root in several colonies. During this struggle, the conflict between various groups in England, clearly appeared. The British ironmasters saw that if the colonists were permitted to continue the establishment of ironworks, they might become serious competitors in the production of pig iron and bar iron, and therefore they demanded higher duties on the importation into England of colonial iron. The iron manufacturers of London, Bristol, Birmingham, Liverpool and other cities who made iron wares, implements and tools from bar iron looked forward to obtaining such supplies of iron as they needed from the colonies at cheap prices. It was natural that they should want the manufacture of all ironwares in the colonies prohibited, thus eliminating all competition there. Woolen merchants and other merchants, together with the shipping and mercantile interests in general, desired that colonial iron be admitted duty free into Great Britain, since British merchandise could be exchanged for it, and commerce given an impetus. The strife between the groups continued even after the Iron Act of 1750 was passed by Parliament, and until the Revolution severed the ties between the two countries.[1]

During this period, England could not supply herself with all the iron needed for her growing iron manufactures and therefore large amounts were received from Sweden and other foreign countries.[2] The accession of George, Elector of Hanover and a member of the anti-Swedish league to the throne of England in 1715 brought the Northern War which was going on at this time close to Britain. It was shown in Parliament that the treaty of mutual defense made between Great Britain and Sweden in 1700 had been continually violated and that British Jacobites were being well received in Sweden. News came that Sweden was inciting a rebellion in England and that troops were sent to support the uprising.[3] While attempts were made to solve the problems which had arisen, no satisfactory agreement could be reached, and George I, authorized by Parliament, prohibited all trade with Sweden in March, 1717.[4]

This act of Parliament cut off the Swedish supply of bar iron, with the result that many engaged in the manufacture of ironwares were thrown out of employment. Petitions were showered upon Parliament by the ironmongers, iron manufacturers, smiths, merchants, dealers and traders in ironware of London, Birmingham, Bristol and other cities asking that commerce with Sweden be restored.[5] Other groups, however, were active, especially the merchants and traders with America who asked that the colonies be encouraged by reduced duties and even bounties to produce iron for the mother country.[6]

The colonial agents attempted to secure aid of some sort for the developing colonial iron industry. Others, like Joshua Gee, one of the mortgagees to whom William Penn pledged his lands in Pennsylvania, was a persistent petitioner to the Board of Trade. Gee, who became a famous ironmaster in Shrewsbury, was anxious at this time for an opportunity to develop the iron mines of Pennsylvania. He appeared before the Board of Trade and asked that duties be removed from all importations of pig iron and bar iron into England and that premiums be granted on such imports. William Byrd also appeared and plead on behalf of Virginia. The many requests to the Board were usually linked with proposals for allowing or continuing bounties on naval stores and other products.[7]

The attitude of the Board of Trade toward the encouragement of colonial pig and bar iron was favorable. It suggested that bounties be granted amounting to £3 per ton on bar iron and £1 – 10 – 0 on pig and cast iron for a period of twenty years. The Board even suggested that if this was done, the colonists could pay all their taxes and quitrents to the Crown in iron.[8] The problem soon found its way into Parliament.

A Naval Stores Bill which was introduced into the House of Commons contained a clause providing that colonial sow iron and pig iron should be admitted into England free of all duties.[9] Through the influence of the London iron manufacturers, another clause provided that no iron wares of any kind should be made in the colonies. The House of Lords went even further and substituted for this a clause forbidding the col-

James Logan

onists to make slit iron.[10] This in effect, if the measure passed, would have meant that not even a nail could be made in the colonies.

All the clauses relating to iron were dropped from the Naval Stores Bill which was finally adopted in 1722.[11] The clauses regarding colonial iron were removed because of the conflict between the various groups. One writer stated that the clauses restraining manufactures really had killed the bill of 1719, "for the 'better disposed' urged that the measure be dropped for that session."[12] The struggle of 1719 was the first of several that were to be waged between the various groups.

The first iron exported from Pennsylvania to England was

made at Durham Furnace. James Logan, very much interested in the export of iron from the colonies to the mother country, in 1728 sent three tons of pig iron to London, hoping to find a market there for the product of this furnace which had just been put in blast.[13] The duty on colonial pig iron had been very quietly reduced in 1724,[14] when Governor Alexander Spotswood of Virginia began shipping his iron to England.[15] Partly through the efforts of Spotswood and William Byrd a duty of only three shillings and nine pence half-penny per ton was imposed on colonial pig iron against £2 – 1 – 6 on iron entering England from all foreign countries. The high duties still remained, however, on bar iron, no preference being shown on this type of iron from the colonies over that from Sweden and other foreign countries.

The attempt to regulate the colonial iron industry came up from time to time. In 1729, the British forge owners were able to exert a sufficient influence to have Parliament draw up a bill in their favor, providing that all forges in the colonies should be destroyed and no new ones set up. It also provided that sow iron and pig iron should enter England from the colonies free of duty.[16] This would have been greatly to the advantage of British forge owners, for if no forges existed in the colonies, the colonial furnaces would be making pig iron to supply their needs, and with the removal of the duty, they would obtain colonial pig iron, which they would forge into bar iron, at a cheap price. The colonial agents used all their influence in opposing the bill because of the clause meant to destroy the forges in the plantations.[17] They were aided by the ironmongers and manufacturers of iron products of England, who wished to secure a cheap supply of bar iron for their needs from the colonies. Together the two groups by "proper and diligent application" were able to have the bill defeated.[18]

In 1735, the controversy opened again, this time in full fury. The first shipments of bar iron (excepting a few tons sent in 1718) from Pennsylvania, Maryland and Virginia at this time contributed to bring about the renewal of the struggle.[19] High duties laid on British goods by Sweden, which still supplied England with most of her bar iron,[20] the tests made at the navy yards of colonial bar iron which proved satisfactory,[21] and the beginning of a depression in the English iron industry as a whole,[22] must also be added to the reasons for the conflict, which was waged for several years. In the petitions to Parliament and the many pamphlets and counter pamphlets that were written during the period from 1735 to 1738, the interests of the various groups were clearly brought out. The furnace owners complained of the low duties on pig iron from the colonies and asked that they be raised, the forge owners demanded that the low duties on pig iron and the high duties on colonial bar iron should remain, the London, Birmingham and Bristol manufacturers of iron wares asked that the duties on such bar iron be removed and that all secondary iron manufactures in the colonies be suppressed, while the shippers cast their lot with the manufacturers because it was to their interests to send their vessels to the colonies with British manufactured goods, and bring back colonial pig and bar iron.[23]

Parliament considered several bills pertaining to the problem in 1738 and 1739. Each bill provided that pig iron and bar iron from the colonies should enter England duty free. Each also had prohibitory clauses. One bill provided that no slitting mills or steel furnaces should exist in the colonies and no more forges or bloomeries for making bar iron should be erected, although those already in operation were permitted to continue.[24] Another bill provided for the admittance of colonial pig and bar iron free of all duties. No more slitting mills or steel furnaces should be built in the colonies, and a prohibitory duty was placed on all foreign steel shipped from Great Britain and Ireland to the colonies. Another restriction in this bill forbade the exportation of all wrought iron manufactures including nails, tools and implements from one colony to another.[25] These bills, however, did not get very far in Parliament.[26]

Once again, because of the varying interests of the different groups, no legislation resulted from all this controversy. The "jarring interests" as one writer puts it, prevented the passage of the law.[27] Conditions in the iron trade in England began to improve and the attention of the country was turned to the difficulties of the European situation which led to the War of the Austrian Succession.

Many interested in the welfare of the colonies continued their efforts to have the duties removed from importations of colonial iron into England. Thomas Penn was exceedingly active. He wrote to the Pennsylvania ironmaster, Richard Hockley, in 1743, telling him of his efforts and stating that he was still "courting the Iron Powers in Parliament."[28]

The War of the Austrian Succession ended in 1748, but before the ink was dry on the Treaty of Aix-la-Chapelle, another war threatened. Russian and Swedish troops were facing each other on the borders of Finland.[29] With France pledged to Sweden and England pledged under certain conditions to Russia, a renewal of conflict appeared inevitable. The agreement between England's political enemy, France, and her industrial competitor, Sweden; the fear of another great struggle; indignation at the treatment of the British representative at Stockholm who was recalled at this time; and opposition to the recently concluded treaty between Sweden and Denmark,[30] were factors which the groups struggling for British regulation of colonial iron industry could not ignore, and which they used to set the stage for another battle. While Parliament was struggling with these foreign problems, the question of regulation of the colonial iron industry was again introduced in February, 1749-1750.[31]

Parliament considered the problem of the unsatisfactory conditions of trade and commerce with Sweden. Another duty had been added by that country to the already prohibitory duties on English manufactured goods. It was shown anew that only a small amount of British merchandise found its way to Sweden, while Great Britain purchased most of her bar iron from that country, with a resulting unfavorable balance of trade. A parliamentary committee was instructed to draw up a bill favoring the free importation of pig iron and bar iron from the colonies.[32] Through pressure exerted by London manufacturers of ironwares, a clause was added, prohibiting the further erection of slitting mills or steel furnaces in the colonies.[33]

While this bill was pending, many petitions were again sent to Parliament for and against it. The same arguments which had been used years before were again brought out by the ironmasters, forge masters, manufacturers, merchants, and shippers.[34] The House of Lords was bombarded in the same manner. The bill, however, in spite of much controversy and opposition, passed both houses, received royal assent and became a law in April, 1750.[35]

The act provided that after June 24, 1750, all duties on pig iron from the colonies should cease. Bar iron could be imported only into London free of all duties. The duties remained on bar iron shipped from the colonies to all other ports of Great Britain. Colonial bar iron sent to London could not be carried by land or water more than ten miles from the city except to the royal navy yards. Heavy penalties were imposed for violations. To prevent fraud, all colonial iron sent to England had to be stamped with some mark denoting the place where it was made. All pig iron not stamped was liable to the duties to which all foreign iron was subject. All bar iron, likewise, sent to the port of London unstamped, was subject to the duties on such iron from foreign countries.[36]

The second part of the act dealt with mills which turned out products beyond the bar iron stage. After June 24, 1750, no more slitting mills, plating forges, and steel furnaces could be erected in the colonies. Those in existence before this date were allowed to continue. Thus three types of iron works were restricted: (1) slitting mills which produced slit iron for making nails; (2) plating forges where iron was hammered into sheet iron or tin plate iron under tilt hammers; and (3) steel furnaces where small amounts of blister steel were made, which was used chiefly for making tools. For every violation of this part of the act, a penalty of £200 was to be imposed, and the works were to be destroyed.[37]

The colonial governors were required to make returns of the slitting mills, plating forges, and steel furnaces in each colony. According to the report made by Governor Hamilton to the Board of Trade, there were two steel furnaces, one slitting mill, and one plating forge in Pennsylvania.[38] There can be no doubt that all the works of these types in Pennsylvania were not returned. Coventry Steel Works on French Creek, established in 1730,[39] was not included although it was still standing when Acrelius wrote in 1757.[40] Several others, also were not reported. This may have been due to negligence in securing the proper information, or it may have been done deliberately.

The wish that the colonies would supply England with pig iron and bar iron was not fulfilled. Thomas Penn, in a letter to Richard Hockley, soon after the act was passed, stated that while he and others could not get all they wished from Parliament, yet he hoped that what had been obtained would be of great advantage to America in time.[41] From the period of the twenties, small amounts of pig iron had been sent from Penn-

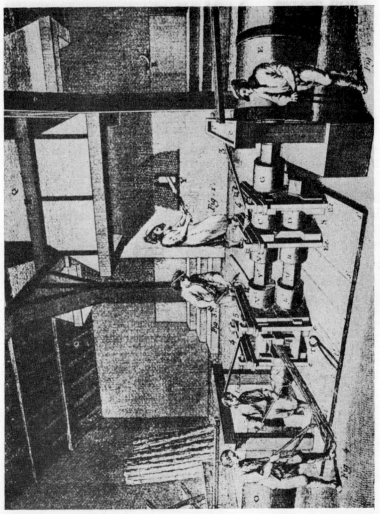

Rolling and Slitting Mill

sylvania to England and also to the West Indies. During the decade before 1750, an average of about 100 tons had been sent to England annually.[42] During the years after the act was passed, this amount increased until in 1755 more than 800 tons were exported.[43] No figures are available for the amount of iron shipped from Pennsylvania for the following years, the reports giving the total from all the colonies. The peak in export of pig iron from all the colonies that shipped iron came in 1771, when a little more than 5,000 tons were exported. From this time on, there was a decline until the Revolution cut off all exports to the mother country.[44] When it is considered that between 40,000 and 50,000 tons of iron were received by England from foreign countries each year,[45] the attempt to encourage colonial pig iron was clearly a failure.

The removal of the duty on bar iron was a much greater concession to the colonists, for the high duty of £2 – 1 – 6 per ton was entirely removed if sent to the port of London. The amount of bar iron sent from Pennsylvania to England before 1750 was insignificant, amounting to only a few tons altogether. After 1750, the amount increased slightly. In 1752, a little more than sixty-four tons were sent to England; in 1753, about 147 tons; in 1754, about 110 tons; and in 1755, only seventy-nine tons.[46] The figures for the years that followed are for all the colonies that shipped bar iron to England. The peak was reached here also, in 1771, when 2,222 tons of bar iron entered England.[47] After 1757, bar iron could be sent duty free to all parts of Great Britain,[48] but this had only a slight effect on the export of it from the colonies. Likewise in 1764, iron was made an "enumerated" commodity in an attempt to have all exports of colonial iron sent to England.[49] But this also failed to produce the desired effect, for the exports of colonial iron were relatively small when compared with the 40,000 to 50,000 tons imported by England from foreign countries. Before the Revolution, those who desired an increase of iron from the colonies attempted to secure an act by Parliament providing for premiums or bounties on colonial iron. In this, however, they were not successful.[50] Thomas Penn and others from Pennsylvania were active in trying to secure such bounties.[51]

The rapid development of the Pennsylvania iron industry has been noted. Before 1750, many furnaces and forges were in operation. During the years from 1750 to 1776, when attempts were made by English manufacturers to obtain colonial iron, a large amount of iron was smelted at the Pennsylvania furnaces and hammered into bar iron at the forges. Since little of this iron left the colony, it is quite evident that the colonists were using it in their rapidly growing manufactures. The numerous calls sent out by the Board of Trade to the governors for a statement of manufactures bears witness to their concern for the industrial progress of the colonies during this period. Owing to the growth of manufactures, the colonists were becoming more and more independent of the mother country and this growing spirit of independence led, after many grievances, inevitably to the Revolution and the severance of ties between the two countries.

The restrictive clauses of the act of 1750 were not seriously regarded in Pennsylvania, and there were many violations of the act. In 1762, Samuel Potts and Company built a steel furnace at Pottsgrove in the Schuylkill Valley and operated it for years and throughout the Revolution.[52] Not very far away in the same river valley, but on the other side of the Schuylkill, Thomas Potts and Company erected a steel furnace and openly advertised their steel as "cheaper than English steel," giving the purchaser the opportunity to try it before buying.[53] This furnace also continued in operation for many years after the Revolution.[54] In 1767, Whitehead Humphreys advertised steel made at his furnace, which was built in 1762,[55] at Seventh Street between Market and Chestnut Streets, Philadelphia, a different location than the two steel furnaces for which a return had been made under the act of 1750.[56]

In 1770, the Provincial Assembly showed its contempt for the law by appropriating £100 to Whitehead Humphreys for encouragement in making steel,[57] and in 1772 it set up a lottery to assist him in the manufacture of that article.[58] After the outbreak of the Revolution, Congress authorized the Board of War to contract with Humphreys for the manufacture of steel. In 1786, the Pennsylvania Assembly granted him a loan to aid him in making steel from bar iron.[59] Many forbidden slitting mills were also erected prior to the Revolution. Joseph and Samuel Potts built a slitting mill before 1775 near Pine Forge,[60] George Ege erected another in the same year,[61] at Charming Forge, while Mark Bird's slitting mill was in operation during the Revolution,[62] and a steel furnace was added soon afterwards.

It can be clearly seen that England's policy for regulating the colonial iron industry was a failure. Above all, the restriction on the erection of certain types of ironworks was a mistake, for no adequate means was adopted to enforce these provisions of the law. It was left to the governor, but no means were given him for putting the law into effect.[63] As was the case with other laws, its violation brought with it an attitude of defiance to the mother country and the prohibition itself was a grievance that irritated the colonists. This act, then, must be included in the numerous causes of the Revolution.[64]

At the outbreak of the Revolution, Great Britain cut off all trade with her American colonies.[65] With the conclusion of peace, the Americans hoped to secure favorable treaties of commerce with the nations of Europe, but they were destined to disappointment.[66] Until 1796, when Jay's treaty went into effect, there was no commercial treaty between the United States and Great Britain.[67] After peace was declared, an act of Parliament passed in April, 1783, renewed commercial intercourse and made it subject to Orders-in-Council issued under authority of that act.[68] Under these Orders, American vessels were allowed to carry to England raw materials, unmanufactured goods, and naval stores from the United States. In return, Great Britain sent manufactured goods. No American ships, however, were permitted to enter any of Great Britain's colonial ports.[69] All proposals for a treaty were rejected until Jay's unsatisfactory treaty was signed.[70]

During the period from the end of the Revolution until the Constitution of the United States went into effect, several of the American states attempted to retaliate against Great Britain for the exclusion of their ships from the West Indies. Discriminating tariff and tonnage duties were imposed. These attempts failed because no uniform and concerted action could be taken.[71] The tariff system of Pennsylvania is an example of the discriminatory tariffs, and was the first attempt in this country to protect iron as well as iron manufactures by means of high duties.

In 1780, while the Revolution was still in progress, Pennsylvania passed a tariff measure which placed low specific duties on wines, liquors, molasses, sugar, coffee, and tea. On all other imports there was levied an *ad valorem* duty of one per cent.[72] Two years later the rates were doubled and the proceeds of the second tariff used for the defense of the Delaware River.[73] Shortly after peace came, the act of 1782 was repealed,[74] leaving the act of 1780 still operative. Thus, most imports paid a duty of one per cent. In 1784, additional *ad valorem* duties of one and one-half per cent were imposed on most imports.[75] Early in 1785, careful provisions were made for the collection of duties.[76] These tariffs were all low and levied for revenue only.

The first Pennsylvania tariff designed to be protective was passed in 1785.[77] In March of that year, a bill to "protect the manufactures" of the state, which levied duties on a large number of articles, among them manufactures of iron and steel, was read in the Assembly for the second time, debated by paragraphs, and then ordered printed for public consideration.[78] A town meeting called in Philadelphia, met on June 2, 1785.[79] A committee was appointed to consider the question of the tariff. It finally declared, that relief from oppressions under which American trade and manufactures labored could come only by a grant to Congress of full powers over the commerce of the United States, and that prohibitive duties should be placed on all foreign manufactures that interfered with domestic industry.[80] A committee was appointed to petition the Pennsylvania Assembly to that effect, and to correspond with committees appointed elsewhere for similar purposes.[81]

A tariff bill, which had been placed before the people of Pennsylvania, was passed by the Assembly in September, 1785.[82] It provided for an increase in duties, especially on iron and steel, and discriminating tonnage duties on ships of countries not having treaties of commerce with the United States. This act "to encourage the manufactures of this state by laying additional duties on certain manufactures which interfere with them," imposed high specific duties on more than forty articles. In addition to the two and one-half per cent *ad valorem* duties levied by the acts of 1780 and 1784, new duties of ten per cent *ad valorem* were imposed on cast iron and British steel as well as on slit iron, nail rods and sheet iron.[83] On all steel, except British, duties of five per cent in addition were imposed, making a total of seventeen and one-half per cent.[84] High duties were placed on steel because the state was attempting to encourage its manufacture, several furnaces being built about this time. The act of 1785 was slightly altered and amended by four subsequent acts,[85] but the protective features were not disturbed. A comparison of this tariff with the first United States tariff will immediately suggest the origin of the latter.

Because of the need for large quantities of bar iron for the growing manufactures of the state, no duties were imposed upon such imports under the tariff of 1785. The forge masters were disappointed and sent a petition to the Assembly bearing 773 names, asking that duties be levied on all importations of bar iron.[86] A clause in the bill to "Encourage and Protect the Manufactures of the State" provided that a duty of £2 – 0 – 0 a ton be imposed upon all foreign bar iron brought into the state. The clause, however, was defeated, and bar iron remained unprotected.[87]

The same year in which Pennsylvania passed her first protective tariff, England passed an act prohibiting the exportation of tools, machinery, engines, models, or plans of machines used in the iron industry, to any foreign country. Heavy penalties were also laid for enticing English workmen employed in iron and steel manufacture.[88] Earlier acts, applied to cotton and other industries, had preceded this.[89] The act passed in 1785, however, was not directed solely against the United States, but at Europe, especially Germany,[90] although naturally it operated against the United States.

Pennsylvania enacted similar legislation in 1788. An act to encourage and protect the manufactures of the state provided that no machine, engine, tool, press, or implement used in the woolen, cotton, linen, silk, iron or steel manufactures should be exported to any country outside the United States. Anyone seducing or encouraging any skilled workers in these manufactures to leave the state was liable to heavy penalties.[91] This law was passed at the time when the spirit of industrialism was seizing the minds of many leaders, who were advocating the development of American manufactures in order to become completely independent of other nations.

The policy of Great Britain in regard to importations of pig iron and bar iron from the United States after 1783 was actually a continuation of the colonial policy in respect to those products. Under the Orders-in-Council, pig iron, bar iron, and many types of naval stores produced in the United States could be imported into Great Britain on the payment of the same duties as on the same sort of goods imported from any British possession in America. Thus pig iron and bar iron

were admitted free of duty, although the very high duties on
iron from all other foreign countries into England were continued.[92]

The Jay treaty went into effect in 1796. It was unsatisfactory to the United States for many reasons, especially because
American ships were still excluded from the West Indies.[93]
The general stipulation in the treaty respecting duties was
that those on American articles should not be higher than on
like articles from any other foreign country.[94] The treaty,
however, provided also that further duties might be imposed
adequate to countervail (but not equal to) the difference of
duty on the importation of European and Asiatic goods into
the United States in British and American vessels.[95] The object of Great Britain in retaining such a right, was, according
to Rufus King, United States minister to Great Britain, not
to check American trade, but to protect British navigation on
the same principle that American at that time protected her.[96]
The tariff act passed by Great Britain in 1797 for putting into
execution some of the provisions of the Jay treaty was exceedingly liberal toward the United States.[97]

It seems remarkable that Great Britain did not take advantage of the terms of the treaty to impose high duties on all
importations from the United States. By the act of 1797, naval
stores of many kinds, wood, staves, pig iron and bar iron,
were put on the same footing as if imported from British colonies, and thus were admitted duty free if carried in British
ships (with certificate).[98] When imported in American ships
(without certificate), a countervailing duty, which was nominal, was imposed.[99] On pig iron and bar iron this was "ten
per cent on the amount of the duties of customs payable on the
said articles imported from any British colony or plantation
in America, when not accompanied by the certificates required
by law."[100] Since iron was imported from the British colonies
in British ships duty free (foreign ships being prohibited from
carrying colonial trade), the duty was figured on the basis of
that paid on iron from foreign countries when imported into
Great Britain. The duty on pig iron from foreign countries
was now 6/0 1/2d; ten per cent of this was estimated to be
6 1/2d.[101] The duty on bar iron from foreign countries in 1797
was £3 – 4 – 7; ten per cent of this was 6/5 1/2d.[102] Thus even
when carried in American ships, the countervailing duties on

iron were slight compared to the duties imposed on foreign iron.[103] During the war, small convoy duties were charged on all importations.[104]

The question arises as to why Great Britain permitted iron as well as other materials from the United States to enter her ports either duty free, or on payment of a very low duty. The question is more significant when we consider that the United States was at this time imposing fairly high duties on iron and iron manufactures from Great Britain and other foreign

Oak Pinion and Gears, Hay Creek Forge

countries.[105] The answer is that England simply continued her colonial policy, begun in 1750. The commodities given preference consisted of raw and semi-raw materials, and were of importance to the manufactures, navigation, navy and marine of England. Even with this encouragement, the amount of iron sent to England was very small,[106] the Americans using most of their iron in their own rapidly growing manufactures. Thus the English ironmasters could not complain of competition. On the other hand, the exceedingly high duties on Russian, Swedish, and other foreign iron remained,[107] for two reasons: (1) to provide a revenue; and (2) to prevent competition with English iron.

The first tariff law of the United States went into effect on July 4, 1789.[108] Its three objects were: (1) the support of the government; (2) the discharge of the public debt; and (3) "the encouragement and protection of manufactures." Specific duties were laid on many articles; *ad valorem* duties, on others. The act laid a higher duty on slit and rolled iron, castings, steel and nails, than upon most other articles. On pig iron and bar iron (cast iron), the rate was seven and one-half per cent.[109]

Alexander Hamilton stated in his Report on Manufactures in 1791, that manufactures of iron, though generally understood to be extensive, were found to be much more so than generally supposed. Many ironworks had recently been built and were more profitable than they had formerly been. The price of bar iron had been increased because of the great demand for it in iron manufactures. He recommended special encouragement to the industry by means of increased duties on all types of foreign iron. He also raised the question of the expediency of permitting the importation of pig iron and bar iron duty free. He thought that it would be favorable to the manufactures of iron, but believed it might interfere with the production of domestic pig and bar iron. However, on account of the high prices obtained for bar iron, it was probable that the free admission of foreign iron "would not be inconsistent with an adequate profit to the proprietors of iron works." He also thought that an increase of bar iron from foreign countries by removing the duties, would result in the increase of iron manufactures, and thus the price of bar iron would not decline. He advised caution and suggested that such a measure ought "to be contemplated subject to the lights of further experience, rather than immediately adopted."[110]

The sections of the tariff act of 1789 pertaining to iron and steel were passed with little difficulty.[111] In 1792, the duties on iron and iron manufactures were raised to ten per cent,[112] and two years later an additional duty of five per cent was levied.[113] All duties imposed by the act of 1789 were subject to a discount of ten per cent if the goods were imported in vessels owned by American citizens.[114] By the act of 1794, an additional ten per cent was added to the duties if imported in foreign vessels.[115]

Before the tariff of 1794 was passed, opposition to the duties on bar iron was expressed by the manufacturers of iron products. In a report from a committee appointed to consider "the propriety of remitting the duty on bar iron in certain cases," it was proposed:

> "That a regulation respecting the duty on bar iron would conduce to the promotion of the growing manufactures of the United States and might tend to prevent a monopoly in the hands of the makers of an article essential to agriculture, improvement and manufacture."[116]

It was suggested that when bar iron should exceed a certain price, to be determined later on the basis of current prices, the President of the United States, by proclamation, should remove all duties on bar iron for a period of two years. A few days later, the ironmasters of Pennsylvania sent a petition to Congress asking that the duties on the importation of cast iron and bar iron from foreign countries be continued, or that Congress give encouragement to the erection and improvement of furnaces and forges.[117] Shortly afterwards the tariff act of 1794 was passed,[118] which reflected the influence of the ironmasters. From 1794 until 1812, the duty of fifteen per cent *ad valorem* on iron and iron manufactures remained. Thus the last two decades of the century witnessed the opening of the tariff issue and its attendant problems, which were to play an important part in American life, politically as well as industrially during the years that were to follow.

FOOTNOTES

[1] See Arthur C. Bining, *British Regulation of the Colonial Iron Industry* for a monograph on this subject.

[2] Almost 20,000 tons were received from Sweden in 1715, and 17,000 more from other foreign countries. All of this consisted of bar iron, no pig iron being imported, and paid a duty of £2 – 1 – 6 per ton if imported in British vessels, and £2 – 10 – 10 in foreign vessels. Harry Scrivenor, *History of the Iron Trade*, Appendix, p. 326.

[3] *Journals of the House of Commons*, XVIII, 474, 726 ff, 750.

[4] *Ibid.* XVIII, 477, 478, 480, 481, 482, 486; 3 George I, c. 1.

[5] *Journals of the House of Commons*, XVIII, 691, 745, 746, 747, 749.

[6] *Ibid.*, XIX, 116.

[7] Board of Trade Journals, XXVI, 183, 194, 195, 205, 208, 209, 212, 215, 220, 221. Board of Trade Papers: Plantations General, IX, K78, K79, K134, K136, 1.24.

[8] *Calendar of State Papers: America and West Indies*, 1716-1717. No. 515.

[9] *Journals of the House of Commons*, XXII, 851; Board of Trade Papers: Plantations General, X, L22.

[10] *Letter to a Member of Parliament Concerning the Naval Store Bill*, pp. 12-13.

[11] *Journals of the House of Commons*, XIX, 281, 316, 669, 704; 8 George I, c. 12.

FOOTNOTES—Continued

12 *Letter to a Member of Parliament Concerning the Naval Store Bill*, p. 12; David MacPherson, *Annals of Commerce*, III, 72.

13 James Logan to Nehemiah Champion, 6th November, 1728, Parchment Logan Letter Book, p. 556.

14 11 George I, c. 1.

15 Board of Trade Journals, XXVI, 217, XXXI, 342-343.

16 Board of Trade Papers: Plantations General, XI, M1.

17 *New Jersey Archives*, First Series, XI, 177-183.

18 Jonathan Belcher to Governor Talcott, May 20, 1729, Talcott Papers, Connecticut Historical Society, *Collections*, IV, 167-169; Governor Talcott to Jeremy Drummer, February 22, 1730-1731, *Ibid*. IV, 220.

19 Harry Scrivenor, *History of the Iron Trade*, Appendix, p. 331.

20 Report of Committee of House of Commons, Penn MSS. XIII, Document 27.

21 *Journals of the House of Commons*, XXII, p. 850; Penn MSS. XIII, Document 77.

22 Report of Committee of House of Commons, Penn MSS. XIII, Document 45.

23 *Journals of the House of Commons*, XXII, 850 ff; Penn MSS. XIII, Documents 1-45.

24 *Ibid.* Document 43; *Journals of the House of Commons*, XXIII, 187, 188.

25 Penn MSS. XIII, Documents 35-39.

26 *Journals of the House of Commons*, XXIII, 172.

27 Adam Anderson, *Origin of Commerce*, III, 218; David MacPherson *Annals of Commerce*, III, 215.

28 Thomas Penn to Richard Hockley, August 19, 1743, Pennsylvania Miscellaneous Papers: Penn and Baltimore, Penn Family, 1740-1756, Document 41.

29 C. H. Firth and J. F. Chance, *Diplomatic Relations of England with the North of Europe*, p. 3.

30 *Parliamentary History of England*, XIV, 494 ff, 822 ff.

31 *Journals of the House of Commons*, XXV, 979. Penn MSS., XIII, 75.

32 *Journals of the House of Commons*, XXV, 979.

33 *Ibid.*, XXV, 986.

34 See Arthur C. Bining, *British Regulation of the Colonial Iron Industry*, Chap. IV.

35 *Journals of the House of Commons*, XXV, 1115.

36 23 George II, c. 29, Sections 1, 2, 7.

37 *Ibid.* Section 9.

38 *Pennsylvania Archives*, First Series, II, 52 ff Board of Trade Journals, LIX, p. 10; Board of Trade Papers. Proprieties, LXV, V72, V73.

39 B. Potts MSS. II Coventry, 1728 (1728-1734) pp. 214 ff. passim. *Ibid.* III, 1730 (1731-1732) pp. 132 ff. passim.

40 Israel Acrelius, *History of New Sweden*, p. 168.

41 Thomas Penn to Richard Hockley, July 1, 1750, Pennsylvania Miscellaneous Papers, Penn and Baltimore: Penn Family, 1740-1756, p. 131.

42 Harry Scrivenor, *History of the Iron Trade*, Appendix, pp. 336-337.

43 *Ibid.* p. 340.

44 *Ibid.* p. 343.

45 *Ibid.* pp. 343-344. Only bar iron was obtained from foreign countries, the duties on pig iron being too high to permit foreign countries to send it to England.

46 *Ibid.* p. 340.

47 *Ibid.* p. 343.

48 30 George II, c. 16.

49 4 George III, c. 15.

50 Board of Trade Papers: Plantations General; S 49, S 50, S 51, S 52 Board of Trade Journals, LXIII, 117-118.

51 Thomas Penn to Lord Sterling, April 9, 1765, Penn Letter Book, VIII, 241-242.

52 B. Potts MSS., XV, Pottsgrove, 1758 (1758-1769) p. 219; *Ibid.* XIX, Pottsgrove, 1765 (1764-1770), pp. 2, 34, 42, passim.

53 *Pennsylvanishe Staatesbote*, November 19, 1764. This undoubtedly was the furnace built in 1732, which was not reported in 1750, possibly because it was not in operation at that time.

54 East Nantmeal Township Rates, Chester County, 1780 on, *Pennsylvania Archives*, Third Series, XII, 342.

FOOTNOTES—Continued

55 In his appeal to the Assembly for aid in 1770, Humphreys stated that he had been making steel for eight years. *Votes of the Assembly*, VI, 200.

56 *Pennsylvania Chronicle*, December 7, 1767, October 22, 1770, January 21, 1771, November 18, 1771.

57 *Votes of the Assembly*, VI, 228.

58 *Pennsylvania Chronicle*, October 24, 1772; *Pennsylvania Packet*, October 26, 1772, *Pennsylvania Packet Supplement*, January 18, 1773.

59 *Independent Gazette*, April 15, 1786, *Pennsylvania Journal*, April 12, 1786, *Pennsylvania Statutes at Large*, XII, c. 1228.

60 Potts MSS. LXX, Pine Forge, 1774 (1744-1781) pp. 78, 109, 130, 170, 188, 249.

61 Charming Forge Account Book, 1775.

62 Potts MSS. LXXIII, Pine Forge, 1781 (1781-1782) pp. 10, 18.

63 This was clearly brought out by a bill discussed by Parliament. *Some Considerations on the Bill to Encourage the Importation of Pig and Bar Iron from America*, Penn MSS. XIII, Document 75.

64 23 George II, c. 29.

65 16 George III, c. 5.

66 F. Wharton, *Diplomatic Correspondence of the American Revolution*, VI, 802.

67 W. M. Malloy, *Treaties, Conventions, International Acts*, etc., I, 590 ff.

68 23 George III, c. 39; *Acts of the Privy Council, Colonial Series*, 1766-1783, pp. 527 ff.

69 *Ibid.* pp. 529 ff.

70 W. C. Ford (Ed.) *Report of a Committee of the Lords of the Privy Council*, pp. 64-76. E. C. Burnet, "London Merchants on American Trade, 1783," *American Historical Review*, XVIII, 769.

71 S. F. Bemis, *Jay's Treaty*, pp. 24 ff.

72 *Pennsylvania Statutes at Large*, X, c. 925.

73 *Ibid.* X, c. 965.

74 *Ibid.* XI, c. 1032.

75 *Ibid.* XI, c. 1076.

76 *Ibid.* XII, c. 1157.

77 *Ibid.* XII, c. 1188.

78 *Pennsylvania Packet and Daily Advertiser*, May 13, 1785.

79 George Bancroft, *History of the Formation of the Constitution of the United States*, 1, p. 187.

80 *Pennsylvania Packet and Daily Advertiser*, June 2, 1785.

81 *Ibid.* Sept. 22, 1785; *Pennsylvania Gazette*, October 5, 1785.

82 *Pennsylvania Statutes at Large*, XII, c. 1188.

83 *Ibid.* section 1.

84 *Ibid.* section 4.

85 *Ibid.* XII, XIII, c. 1198, 1226, 1276, 1346.

86 *Minutes of the Assembly*, 11th Assembly, 1st Session, p. 62.

87 *Ibid.*, 11th Assembly, 1st Session, pp. 107-108.

88 25 George III, c. 67.

89 14 George III, c. 71.

90 *Debates and Proceedings of the House of Commons*, III, 543 ff.

91 *Pennsylvania Statutes at Large*, XIII, c. 1347.

92 W. C. Ford (Ed.), *Report of a Committee of the Lords of the Privy Council*, pp. 6 ff.

93 S. F. Bemis, *Jay's Treaty*, pp. 24 ff.

94 *Jay Treaty*, Article 13; *American State Papers, Foreign Relations*, II, 112.

95 *Ibid.*, II, 112 f. *Jay Treaty*, Article 15.

96 *American State Papers, Foreign Relations*, II, 112 ff.

97 37 George III, c. 97.

98 *Ibid.*, Sections 3, 11, 12, 13.

99 *Ibid.*, Section 3.

100 *Ibid.*, Section 11.

101 E. J. Mascall, *Book of Customs*, p. 56.

102 *Ibid.*, p. 55.

103 37 George III, c. 97, Section 3.

104 E. J. Mascall, *Book of Customs*, pp. 1, (note) 55, 56.

FOOTNOTES—Concluded

[105] Tariff Acts of 1789, 1792, 1794. *House Documents*, 55th Congress, 2nd Session, No. 562, pp. 9 ff., 29 ff., 34 ff.

[106] *American State Papers, Foreign Relations*, II, 112.

[107] E. J. Mascall, *Book of Customs*, p. 55.

[108] Tariff Act of 1789, *House Documents*, 55th Congress, 2nd Session, No. 562, pp. 9-14.

[109] *Ibid.* p. 10.

[110] *American State Papers: Finance*, I, 138.

[111] William Maclay, *Journal*, p. 57; *Annals of Congress*, 1st Congress, 1789-1791, I, 105 ff.

[112] Tariff Act of 1792, *House Documents*, 55th Congress, 2nd Session, No. 562, pp. 29 ff.

[113] Tariff Act of 1794, *Ibid.* pp. 34 ff.

[114] Tariff Act of 1789, *Ibid.* p. 11.

[115] Tariff Act of 1794, *Ibid.* p. 35.

[116] *American State Papers: Finance*, I, 275.

[117] *Annals of Congress*, 3rd Congress, 1st Session 1793-1795, pp. 523-524.

[118] Tariff Act of 1794, *House Documents*, 55th Congress, 2nd Session, No 562, pp. 34 ff.

CHAPTER IX

THE PROGRESS OF THE IRON INDUSTRY

THE history of the Pennsylvania iron industry during the years from 1716, when the first bloomery forge was built, to 1800, when the industry had achieved remarkable development, is not so much a record of alternate periods of prosperity and depression, although there were such periods, as it is the story of the outstanding successes of many ironmasters and the tragic failure of others. In the new country where there was a demand for iron for various purposes, the establishment of ironworks during the early years was limited by a lack of capital to erect such works and by the inability to obtain ready money to carry on such enterprises. A shortage of working capital was the chief obstacle to the development of the early iron industry.[1] Few men who wished to engage in the manufacture of iron had the necessary money to do so. It was for this reason that groups of men usually combined to establish and finance ironworks.

In an undeveloped country, suffering from a continual lack of capital, a relatively large amount of money was required to build and equip a plant for producing iron. An idea of the cost of eighteenth century ironworks may be derived from a few illustrations of sales and evaluations. Governor Keith stated that his works erected in 1724 had cost him £4,000,[2] although an enemy replied that scarcely anyone would accept the works as a gift, if obliged to carry them on.[3] The Durham Iron Works, which consisted of a furnace and three forges, together with many necessary buildings and the plantation, was sold by the Commissioner of Forfeited Estates in 1779 for £12,800.[4] In 1786, a three-quarter interest in the Carlisle or Boiling Springs Iron Works was sold to Michael Ege for £1,714 – 5 – 9.[5] About the same time, one-half of Mount Hope Furnace and Hopewell Forges and plantations, together with one-sixth of

169

the ore banks, sold by Henry B. Grubb to Robert Coleman, brought £11,100,[6] while Peter Grubb sold to Robert Coleman a one-sixth interest in Cornwall Furnace, buildings and lands, together with a third of Hopewell Forges and lands, for £8,500 in gold.[7] In 1797, a quarter interest in the Westmoreland Iron Works in the western part of the state brought £1,850.[8] In 1786, Henry Drinker estimated that an iron plantation, including furnace, forges, gristmill and sawmill, and all equipment together with the necessary amount of land, would be cheap at £6,000 and dear at £12,000.[9]

Plantations having simply forges on them also sold at relatively high prices. The estate of William Bird upon which were three forges was valued at £13,000 in 1764.[10] Henry William Stiegel paid £3,132 for a half interest in the forge and eighty-eight acres of land on Tulpehocken Creek which became known as Charming Forge.[11] In 1799, Davies Old sold to Cyrus Jacobs, Poole Forge and 783 acres of land in Lancaster county for £10,000.[12] During the last decade of the century, Tench Coxe made some very modest estimates of the cost of the smaller types of ironworks. He believed a steel furnace could be constructed for $3,000, a slitting mill for $5,000, a tilt hammer for $1,000, a nailery for $500 and a blacksmith shop for about the same figure.[13]

In order to build and operate ironworks, companies which were really partnerships were formed. Colebrookdale Furnace, the first blast furnace in the province, erected in 1720, was owned by a company of which Thomas Rutter, the pioneer ironmaster, held the largest interest. In 1731, this works was owned by Nathaniel French (three-twelfths interest), Alexander Wooddrop (three-twelfths), Samuel Preston (one-twelfth), John Leacock (one-twelfth), George Mifflin (one-twelfth), T. Potts and G. Boon (one-twelfth).[14] Pool Forge, not far away from Colebrookdale, was built in 1725-26 by a company consisting of James Lewis, Thomas Marke, Alexander Wooddrop, F. Rawle and Robert R. Griffiths.[15] The famous Durham Iron Works was built in 1727 by a company consisting of twelve partners, including James Logan.[16] The ill-fated Alliance Iron Works, built in the western part of the state in 1789-90, was controlled by three partners, William Turnbull, Peter Marmie and John Holker during the early years of its career.[17] Westmoreland Iron Works was established by a company consisting of Martin Dubbs, Charles Marquedant, John Gloninger and Christopher Lobingier.[18] Almost all the ironworks with few exceptions were built and controlled by a number of partners, one of whom lived at the mansion house on the iron plantation carrying on the enterprise and being responsible for it to the company.

The capital invested in ironworks came from many sources. In contrast, however, to the early development of the Virginia iron industry in the eighteenth century,[19] English capital played but a very small part in the establishment of industry in Pennsylvania. In the years following the Revolution, many attempts were made to obtain European capital. One of the most enthusiastic in trying to interest Dutch financiers to invest money in American industries was James Wilson. In 1785 he tried to obtain a loan of 500,000 florins from Holland for his enterprises in the iron trade, in partnership with his brother-in-law Mark Bird. He offered as security for the loan "a very extensive system of works consisting of rolling and slitting mills, grist mills, saw mills, and a forge on the River Delaware, 30 miles from the city" [Philadelphia].[20] He attempted to point out to the Dutch the opportunities they might find by lending money to the Americans[21] to promote their growing manufactures.[22] His efforts at this time were not very successful, but not long afterwards he obtained money from the Holland Land Company for his many land speculations.[23] When he died in 1797, his estate was indebted to the Dutch Company for $46,578.38.[24]

Much of the capital obtained for establishing ironworks was diverted from mercantile channels. Of the twelve original partners of the Durham Iron Works mentioned above, six were merchants.[25] Several members of the Potts family had stores in Philadelphia as well as iron plantations in the Schuylkill Valley.[26] Turnbull and Marmie in the West were merchants, who attempted to increase their fortunes by manufacturing iron.[27] The number of merchants who invested in ironworks was relatively large. As the industry expanded westward, some of the capital invested came from the mercantile interests of the older East.

A large part of the capital that found its way into the iron industry was invested by men who represented a variety of occupations. Many blacksmiths, ironworkers, farmers, and others, who through constant toil, thrift and perseverance were able to accumulate sums of money often invested their savings in ironworks. Thomas Rutter, the blacksmith; Anthony Morris, the brewer; Isaac Meason, the farmer who became a distinguished judge; Robert Coleman, the clerk; George Taylor, also a furnace clerk; and James Old, the ironworker, represent a few of the occupations of men who held shares in ironworks, many of whom became prominent ironmasters.

Mansion House, Pine Grove Furnace

Because of the lack of capital and ready money, ironmasters could give little credit to their customers. Henry Drinker wrote in answer to an application for credit in 1786: "As to terms of payment, ironmasters are generally needy and seldom qualified to give long credits. If payment is delayed until the whole iron is delivered, it seems to be as much as I can agree to, and should suppose it would accommodate your views."[28] The difficulty in securing cash for their iron caused the failure of many ironmasters.

Throughout the eighteenth century, all the colonies, and later the states, were confronted with the problem of securing a circulating medium large enough to satisfy existing needs. As trade expanded and manufactures, especially that of iron, began to develop, the problem became more acute. While English, French, German, Dutch, Portuguese, and Arabian gold and silver coins were current, and although the legislature issued bills of credit and other paper currency from time to time, there was continually a scarcity of money.[29] Thus barter, common from the time of the Duke of York's settlement,[30] continued to a greater or lesser extent for more than a hundred years.

Bills of credit were issued in Pennsylvania, as in the other colonies, either on loan, land or otherwise; to meet some sudden emergency; or to furnish aid for the ordinary operations of the government. Following many petitions from the residents of Bucks, Chester and other counties in 1722-3, the legislature passed on an act for emitting £15,000 in bills of credit.[31] In December of the same year a new issue of £30,000 was authorized.[32] The laws authorizing these issues provided that the bills were to be loaned out on land security, or plate of treble value deposited at the loan office, at five per cent interest. They were made a legal tender. Annual payments were to be made by the borrowers of one-eighth the principal and the interest. These issues were renewed by the Assembly and many new issues were also authorized. Benjamin Franklin played a part in securing this legislation,[33] and especially in having some of the issues made irredeemable.[34] Later issues provided money for the use of the king and for the support of the Pennsylvania government.

The bills were extensively counterfeited. A law was passed in 1729 imposing heavy penalties for counterfeiting the bills, which included the loss of both ears, a fine of £100 and a payment of double the loss sustained by the defrauded.[35] In the case of inability to pay, the offender was sold into service for seven years. It was discovered that much of the fraudulent paper had been made in Ireland and exported to Pennsylvania.[36] In 1756, the penalty for counterfeiting was changed to death.[37]

Up to the time of the Revolution, laws had been passed calling in the bills emitted during the dominion of Great Britain. By a law passed in 1778, all bills emitted prior to April 19, 1775, were declared to be no longer legal tender and for a short period were to be received in payment of taxes and in exchange for the later notes.[38] Before this, in 1776, an act had been passed which made all bills of credit issued by the Continental Congress and by the state of Pennsylvania legal tender.[39] Several issues of bills of credit were authorized after the Revolution and were redeemed by acts passed before the end of the century. A final act was passed in 1805 which provided that all bills outstanding and not paid into the Treasury before January, 1806 should be forever irredeemable.[40]

During the colonial period many attempts were made by the Board of Trade to fix rates for the various gold and silver coins,[41] but without much success. In 1743, seventy-five Philadelphia merchants made an agreement whereby English guineas, French guineas, Portuguese coins, Dutch or Guinea ducats, German carolines, Arabian chequins, French pistoles, Spanish pistoles and other gold and silver coins were to be received at a definite and fixed value. This agreement, published in the newspapers, was in force for three years.[42] From time to time, tables of foreign coins and their value were used.[43] The medley of foreign gold and silver coins remained in circulation well into the beginning of the nineteenth century.[44]

Throughout the entire period, payment was often made by means of exchanges of goods. In many places this was restricted to American manufactures.[45] In the more remote regions, the chief mediums of exchange were peltries, furs, hides, and skins.[46] Whiskey was a common medium of exchange in the West during the latter part of the century,[47] although iron was exchanged for beef, pork, linen, linsey and sugar, as well as for whiskey.[48]

In return for much of the iron sold, therefore, the ironmasters received merchandise, provisions and supplies, which could be used on the plantations. For example, John Taylor of Sarum Iron Works sent a shipment of bar iron to Philadelphia to be sent to Boston. He ordered "the returns to be made in Oil, Loaf Sugar and Rum or such other goods" which he could use on his plantation.[49] The ironmasters in the Schuylkill Valley and neighboring regions exchanged some of their iron for needed commodities with the merchants of Philadelphia. In the central and western sections, exchange was made with merchants in the boroughs and large villages.[50]

When currency was received by the ironmasters, whether in part payment or entire payment for iron, a variety of money usually changed hands. The following illustration is one of hundreds. For a small amount of iron sold at Colebrookdale in 1739, Thomas Potts received a note for £6, three ounces of gold at thirty-five shillings per ounce, paper money valued at £33 – 15 – 0, and copper pence to the amount of £2 – 10 – 0.[51] Even in Philadelphia, merchants received a variety of coins. In 1786, Henry Drinker received the following in payment for a bill:

2 Doubloons or 8 Spanish Pistoles.... (28/–)	£11 –	4 – 0
1 whole, 30 half and 1 quarter Joe.... (120/–)	£97 –	10 – 0
43 whole, 2 half and 1 English guinea. (35/–)	£77 –	8 – 9
5 whole and 1 half Fr. guinea........ (34/6)	£ 9 –	9 – 9
2 whole and 1 half Caroline.......... (35/–)	£ 4 –	7 – 6 [52]

Owing to the lack of ready money, John Hayden the ironmaster of Fairfield Furnace, in the western country, issued notes or bills of credit of his own for one year in denominations from one to ten dollars. He asked the public to accept this currency at the same rate as gold and silver in return for what he had done in providing iron for the western pioneers. "I have spent," he declared, "upwards of a thousand nights at hand labor while others were taking their ease in bed, beating off ice from the wheels and keeping business going; my furnace blows almost without ceasing; metal can be had at all times at reasonable terms."[53] He hoped to secure in this manner £4,000 or £5,000 to buy stock, supplies and provisions for his plantation for the ensuing year. This enterprising ironmaster, however, continued but a few years and then became bankrupt.

The successes and failures of certain individual ironmasters have been discussed in a previous chapter. It remains here to trace the gradual development of the industry as a whole to the time of the Revolution, the impetus given to the production of iron especially munitions during the struggle with the mother country, the increased activity in the period following the Treaty of 1783, the establishment of the industry in the West, and the rapid development of secondary iron manufactures throughout the state. The early period was one of gradual growth. The last decade or more of the century witnessed the establishment of many works in all parts of the state, even in the frontier country of the West. The period of rapid development during the latter part of the century occurred at the same time as the period of prosperity and expansion in the English iron industry, which followed the changes in the technical processes of ironmaking in that country during the Industrial Revolution. The English inventions, in general, however, were not taken over by American ironmasters and the old methods continued for many years after the century closed.

James Logan Fireback

During the early years of the Pennsylvania iron industry, from 1716 when the first bloomery was built, until after the middle of the century, the ironworks were confined largely to eastern Pennsylvania.[54] These years, when pioneer ironmasters such as the Rutters, Potts, Nutts, and others were establishing the industry, were on the whole years of prosperity. There was little competition; good ores, forests and water power were plentiful; and iron for a variety of uses was in demand. Pig iron sold as high as £8 a ton in 1728 at the furnaces. In Philadelphia the price was £10 to £11 a ton. By 1740 it was down to

£5 a ton at the furnaces, and £7 a ton in Philadelphia. There was little change in 1750, although in 1760, it was selling at £7 a ton at the furnaces and £9 a ton in Philadelphia.[55] The greatly increased price of pig iron in Philadelphia compared to the price at the furnace was due to the high cost of transportation.

Hermelin gives the cost of production in 1783, when pig iron was selling at £8 to £9 a ton in Philadelphia, as follows:

At the Warwik [Warwick] blast furnace in Pennsylvania, the figures are . . . for one week's production:

	Pounds	Shillings
45 tons of ore at 10 shil[lings per ton]...	22	10
70 loads of coal at 25 shil[lings per load].	91	
4 tons of limestone at 14 shil[lings per ton]	2	16
Wages for smelters and laborers.........	17	15
Clerical expense, materials and repairs....	4	10
Total	138	11

This makes the cost of production 4 pounds 12 shil[lings] per week for 30 tons of pig iron. [The cost] at the Hopewell blast furnace is about the same, but at the Cornewall [Cornwall] blast furnace slightly less, so that 4½ pounds [per ton] are on an average the cost of production in Pennsylvania.[56]

The price of bar iron varied from time to time. In 1740, bar iron sold in the Philadelphia market at £30 a ton. In 1750, the price was £27 a ton; in 1760, £32 a ton; in 1765, £25 a ton; and in 1772, £28 to £30 a ton.[57]

Hermelin states that during the Revolution the selling price of bar iron doubled, but by 1783 it had returned to normal. His figures on the cost of production are as follows:

	Pounds	Shillings
27 hundred weights pig iron at 4½ pounds Pennsyl[vania] money	6	1½
4½ loads of coal at 100 bushels [per load] at 1 pound, 6 shil[lings]	5	17
Wages for forging	4	
Clerk	1	
Buildings, interest on stock supply and extra expenses	2	
Cost of transportation	1	15
Total	20	13½

At this time, the price of bar iron in Philadelphia was £27 to £30 a ton.[58] In the years that followed prices went higher.[59] From 1789 to the end of the century, pig iron was quoted in the Philadelphia market at £8 a ton, and bar iron from £26 to £27 a ton. Castings brought £22 to £30 a ton, sheet iron, £60 to £65 a ton, and nail rods, £33 a ton.[60]

By the middle of the eighteenth century, many furnaces, bloomeries, refinery forges, and other types of iron works had been built in Pennsylvania and the province was beginning to attain industrial leadership over the other colonies. Acrelius, writing just after the midway point of the century had been passed wrote that "Pennsylvania in regard to its iron works is the most advanced of all the American colonies. When New Jersey is added to it, one can say that from the Delaware, the greatest part of the iron in America is taken."[61] Complying with the provisions of an act of Parliament requiring an account of the bar iron made in the province between Christmas, 1749, and January 5, 1756, Governor Denny reported the output of eight forges and stated that two forges had not yet sent in their records. Whether from lack of knowledge or evasion, Denny did not report all the forges in the province.[62] Besides those he named, others were in operation. The account books and records of Tulpehocken Eisenhammer, Durham Forges, Hopewell Forge, Sarum Forge, and others[63] show clearly that these forges were turning out bar iron at this time.[64] Parliament also required a statement of the amount of pig iron made for the same period, but Governor Denny made no returns.[65]

The amount of iron exported from Pennsylvania down to the middle of the century was small, almost negligible.[66] The bar iron made in the colony was worked into the needed implements and wares which were used in Pennsylvania and the neighboring regions. After 1750, Pennsylvania pig iron was sent to New England forges, while pig iron and bar iron were exported to England and to the West Indies in increasing quantities.[67] While iron was made an enumerated commodity in 1764[68] the law had little effect on the Pennsylvania market since direct exportation was permitted to all parts of America, Asia and Africa.

No accurate record can be given of all the iron exported from Pennsylvania to England. Much pig iron was sent from the Cornwall Works and from other works in the same region, down the Susquehanna River. This iron was exported from points in Maryland.[69] This partly accounts for the relatively large amounts credited in the Custom House Papers to Maryland and Virginia which were always connected in the reports. Records evidently were not kept of the separate shipments from Pennsylvania by way of the Susquehanna River to Maryland.

During the decade of the sixties and through the seventies until the Revolution cut off all exports, Pennsylvania became the leading colony in the export of bar iron to England. From April 5, 1765 to April 5, 1766, Pennsylvania exported 882 tons of bar iron,[70] while the total amount exported by all the colonies did not exceed 1,200 tons. Besides this, during the same period 813 tons of pig iron and about twenty tons of hoops were shipped from Philadelphia.[71] In the records for each of the years 1771, 1772, and 1773, exports of Pennsylvania pig and bar iron are combined. In 1771, a total of 2,358 tons of iron and about forty tons of hoops left Pennsylvania[72] out of a total of 7,500 tons from all the colonies. In 1772, a total of 2,205 tons of iron were exported from Pennsylvania[73] in contrast with 4,690 tons of iron from all the colonies. In 1773, when the colonies exported 3,774 tons of iron, Pennsylvania shipped 1,564 tons of it and about thirty tons of hoops.[74]

Down to the period of the Revolution, the iron industry of Pennsylvania had achieved a normal growth. The wars of the eighteenth century had but slightly influenced its development to this time. Munitions and weapons of war for colonial wars had been brought over from England, although colonial furnaces had occasionally cast cannon and shot.[75] A more settled mode of living; the growth of population, of towns, of wealth to a limited degree; and the resultant need and demand for ironwares, brought about industrial activity from the years following closely upon the Treaty of Utrecht over the period to the beginning of the struggle that led to the independence of the colonies. Although Pennsylvania was one of the last of the thirteen original colonies to be established, she made progress very rapidly in agriculture, commerce and manufacturing.[76]

The development of the iron industry in the colonies made possible the success of the separation from the mother country. If the industry had not reached such a high stage of development, the colonists would have been helpless in the struggle. One of the outstanding factors which led to the successful terminus of the struggle, the importance of which cannot be overestimated, was the colonial iron industry which turned out cannon, shot and munitions for the Patriot troops as well as iron and steel for weapons which were used in the campaigns of the Revolution.

As early as 1770, the Provincial Assembly of Pennsylvania had defied Great Britain's prohibition of the erection of steel furnaces by encouraging steel making.[77] With the outbreak of the struggle, the Provincial Convention recommended to the inhabitants of Pennsylvania the manufacture of steel, iron, wire and tin plate,[78] while the Continental Congress contracted with Whitehead Humphreys for the manufacture of steel which was to be used for making weapons of war.[79]

Most of the furnaces of eastern Pennsylvania made cannon, shot and munitions for the Revolutionary War. Oley Iron Works, Warwick Iron Works, Reading Iron Works, Durham Iron Works, Berkshire Furnace, Codorus Iron Works, Mary Ann Furnace, and others provided the Continental cannoneers with munitions.[80] At Cornwall Iron Works, munitions and salt pans were made. With the supply of salt cut off, the need of salt for salting meats, curing fish, and for various domestic uses, became acute. Salt works were established on the Jersey coast, making salt from sea water. Thousands of salt pans were cast at the furnaces.[81] At the Carlisle Armory, attempts were made to forge cannon from bar iron gads,[82] but without much success.

As was natural, industrial activity, especially the manufacture of munitions and supplies for the army, greatly increased during the war. Not only were iron cannon cast, but brass cannon were also made.[83] Air furnaces were erected during the emergency to cast iron cannon from pig iron or old metal.[84] Boring mills were established.[85] A provincial gun lock factory was set up,[86] and the armory, established about 1761[87] at Carlisle, which later became known as Washingtonburg, turned out swords, pikes, and weapons of various kinds as well as muskets.[88] Contracts were made for arms and muskets in Philadelphia, Lancaster, York, Bedford, and many other places[89] A sheet iron plant was built in Philadelphia by Murray, Griffin and Bullard in 1776 where camp kettles, blaze pans, frying pans, stew pans, tea kettles, and other ironwares could be obtained.[90] Several forbidden plating mills, slitting mills, and steel furnaces were erected.[91]

The years immediately following the Peace of Paris which brought the war to a close and gave the colonists their liberty were years of reconstruction and depression. With the return of prosperity about 1787, the Pennsylvania iron industry wit-

Stove Plate Cast at Durham Iron Works

nessed a remarkable expansion. In 1794, Tench Coxe estimated that "taking into the calculation the extent and number of the existing furnaces and forges of Pennsylvania, the new iron works of the last seven years are equal to one half of all those, which had been erected in the state during and before the year 1787."[92] With but few exceptions, all the older works were in operation, and this period of prosperity and rapid expansion continued until after the end of the century.

During this new industrial era many slitting mills were established. It was estimated in 1794 that the Pennsylvania slitting mills then in operation cut and rolled 1,500 tons of slit iron for making nails.[93] Steel works were also built, and attempts were made to obtain steel by the German method of smelting.[94] Air furnaces were built.[95] Tin plate works sprang up,[96] but attempts at drawing wire which began as early as 1775 were all failures. Until after the close of the century, almost all the wire used came from Europe.

The period after the year 1787 and entering the beginning of the nineteenth was also one of great development in the manufacture of ironwares and secondary iron products of all kinds. A great quantity of iron was used in shipbuilding in the form of spikes, nails, chain plates and rudder iron. From the first days of Quaker settlement, vessels of various types were built along the Delaware. Oldmixon pointed out the extent of the industry at Philadelphia, New Castle and at other ports.[97] At Marcus Hook, shipbuilding yards were supplied with iron from the Sarum Iron Works not far away.[98] Large boats were built even in inland towns such as Reading.[99] Tench Coxe, writing in the last decade of the century, stated that shipbuilding in Philadelphia exceeded that of most parts of the world.[100] For purposes of shipbuilding, Pennsylvania iron was preferred to European because of its pliability and extraordinary toughness.[101]

Long before the Revolution, iron manufactures of all kinds had been developed in different localities. Most secondary iron products were made by blacksmiths. In 1788, there were 214 blacksmiths in Philadelphia alone, while large and small blacksmith shops could be found in all parts of the state.[102] Such communities as Germantown and Lancaster had become noted as manufacturing centers. Lancaster, for instance, before the century closed had twenty-five blacksmiths and whitesmiths, six wheelwrights, four tinners, seven gunsmiths, seven nailmakers besides silversmiths, potters and all types of artisans. In 1786, there were seventeen furnaces, forges, rolling and slitting mills within thirty-nine miles of the town.[103] York likewise was a manufacturing center.[104] In the western towns of Huntingdon, Bedford, Pittsburgh, and Washington, artisans of all kinds could be found, including blacksmiths, whitesmiths, nailors, tinners, wheelwrights, and gunsmiths.[105]

Soon after the Revolution, Pennsylvania iron could be obtained again in the London market. But the amounts of pig iron and bar iron exported from the state were much less than in the years before the break with the mother country. For the year ending September 30, 1792, there was exported from Pennsylvania 797 tons of pig iron, but only twenty-one tons of bar iron, the large amount of iron produced at this time being used in iron manufactures within the state. In addition, for the same year, forty casks of nails, nineteen anchors, sixteen cannon, 740 shot, seventy tons of castings and seven and one half tons of hoops left Philadelphia for different parts of the world.[106] By the last decade of the century the dependence upon Europe for many articles had almost come to an end. Less steel than ever was imported, and nails together with wool and cotton cards were being exported in increasing quantities.[107]

Large amounts of iron from England and from the Baltic countries were imported to be manufactured into iron products. In 1790, St. Petersburg alone sent to the United States 1,288 tons of bar iron and more than forty tons of hoops and nail rods.[108] The bar iron imported from foreign countries during this period is proof of the remarkable development of secondary iron manufactures.

No accurate record of the extent of the Pennsylvania iron industry, and of iron manufacturing, appeared until the census of 1810. At that time there were forty-four blast furnaces, seventy-eight forges, six air furnaces, four bloomeries, eighteen rolling and slitting mills, fifty trip hammers (plating mills), five steel furnaces, and 175 naileries in the state.[109] It is true that a number of these were built after 1800. But the majority of them had been erected and were in operation long before the close of the eighteenth century. By that time the industry had been well established, although its fundamental characteristic was its scattered nature, ironworks and iron manufacturers being found in a large number of sections of the State.

FOOTNOTES

1 Lieutenant-Governor Clarke to the Lords of Trade, June 2, 1738, *Documents Relative to the Colonial History of the State of New York*, VI, 116. J. D. Schoepf, *Travels in the Confederation*, I, 118. *Universal Asylum and Columbian Magazine*, 1791, pp. 169 ff.

2 William Keith, *Just and Plain Vindication*, pp. 7-8.

3 James Logan, *A More Just Vindication*.

4 *Pennsylvania Colonial Records*, XII, 104.

5 Cumberland County Deed Book, H, I, 214 ff.

6 Lancaster County Deed Book, C, III, 514 ff. Muniments of Title, No. 16.

7 Lancaster County Deed Book, pp. 35 ff. Agreement is in *Ibid*. EE, pp. 156 ff.

8 Westmoreland County Deed Book, III. 170-171.
9 Henry Drinker to Richard Blackledge, 10th Month, 4th, 1786, Drinker Letter Book, 1786-1790, p. 84.
10 Inquisition into the Estate of William Bird, Anna Krick MSS.; *Alden's Appeal Record*, Coleman vs. Brooke, p. 441.
11 Berks County Deed Book, IV, 310 ff.
12 Lancaster County Deed Book, 1799.
13 Tench Coxe, *View of the United States*, pp. 384 ff.
14 Potts MSS. Colebrookdale Ledgers, Mrs. T. P. James, *Memorial of Thomas Potts, Jr.*, p. 44.
15 Pool Forge Ledger 1725-1727.
16 James Logan to John Penn, December 6, 1727, Logan Papers, I, Document 89. The partners held the property as tenants in common. The entire interest was divided into sixteen shares and a trust formed for fifty-one years. Philadelphia County Deed Book, G, III, 240 ff. Logan held four shares. James Logan to Nehemiah Champion, 6th November, 1728, Parchment Logan Letter Book, p. 556.
17 Fayette County Deed Book, B, pp. 39-40, 319.
18 Westmoreland County Deed Book, II, 55 ff, 393 ff; III, 54 ff.
19 Alexander Spotswood and his partners, for example, built the first three works in Virginia using English capital. See ".A Progress to the Mines" in J. S. Bassett, *Writings of Colonel William Byrd of Westover in Virginia, Esq.*"
20 James Wilson Papers, III, 24.
21 Many others were trying to secure European capital during this period.
22 James Wilson Papers, III, 26.
23 Van Eeghen Collection, Chest 9, Letter Book, Cazenova to the Four Houses, July 6, 1790, cited in P. D. Evans, *Holland Land Company*, pp. 11, 31-32, 87 ff.
24 *Ibid.* p. 89.
25 Philadelphia County Deed Book, G, III, 240.
26 Mrs. T. P. James, *Memorial of Thomas Potts, Jr.*, p. 104.
27 *Pittsburgh Gazette*, April 12, 1788, November 22, 1788, November 2, 1793.
28 Henry Drinker to Alexander Macomb and William Edgar, 1st 7th Month, 1788, Drinker Letter Book, 1786-1790, p. 293.
29 In the early days of settlement, Pastorius wrote: "We are endeavoring to introduce the *cultivation* of the *vine*, and also the *manufacture of woolen cloths* and *linens*, so as to keep our own money as much as possible in the country; for this reason we have already established *fairs* to be held at stated times, so as to bring the people of different parts together for the purpose of barter and trade and thereby encourage our own industry and prevent our little money from going abroad." F. D. Pastorius, *Province of Pennsylvania*, Historical Society of Pennsylvania, *Memoirs*, IV, Part 2, pp. 83-113.
30 *Upland Court Records*, pp. 59 ff.
31 *Pennsylvania Statutes at Large*, III, c. 261.
32 *Ibid.* III, c. 275.
33 Benjamin Franklin, *A Modest Enquiry into the Nature and Necessity of a Paper Currency.*
34 *Pennsylvania Statutes at Large*, IV, c. 319.
35 *Ibid.*, IV, c. 306.
36 *New York Gazette*, March 13, 1726.
37 *Pennsylvania Statutes at Large*, V, c. 412, section 7.
38 *Ibid.* IX, c. 791.
39 *Ibid.* IX, c. 723.
40 *Ibid.* XVII, c. 2623, section 3.
41 See Catalogue of Papers Relating to Pennsylvania and Delaware, Historical Society of Pennsylvania, *Memoirs*, IV, pp. 231 ff.
42 *Pennsylvania Gazette*, September 16, 1742.
43 John Rowlett's *Tables of Coins, Pittsburgh Gazette*, June 7, 1800.
44 Thomas Cooper, *Some Information Respecting America*, p. 145. Cramer's *Pittsburgh Almanack*, 1812.
45 *Pennsylvania Gazette*, April 27, 1791.
46 J. Doddridge, *Notes on the Settlement and Indian Wars of the Western Parts of Virginia and Pennsylvania*, p. 147; J. D. Schoepf, *Travels in the Confederation*, I, 222.

FOOTNOTES—Continued

47 *Pittsburgh Gazette*, November 2, 1793.

48 *The Union*, December 1, 1800, *Tree of Liberty*, March 13, 1803.

49 Extracts and Memoranda of the Private Papers of John Taylor, Folio, Book A, Cope Collection, p. 138.

50 Such merchants dealt in a variety of goods. Stoves, grates, cast iron wares, and bar iron could usually be obtained at such places of business. For example, see advertisements in *The Union*, December 1, 1800.

51 Potts MSS. Colebrookdale, 1735 (1735-1747), p. 328.

52 Henry Drinker to George Bowne, 10th Month 9th, 1786, Drinker Letter Book, 1786-1790, p. 84.

53 *Pittsburgh Gazette*, January 30, 1801, February 6, 1801, etc. *Tree of Liberty*, March 7, 1801.

54 The iron works up to the close of the French and Indian War in 1763 lay within a radius of seventy miles of Philadelphia. Israel Acrelius, *History of New Sweden*, p. 169.

55 Potts MSS. II. Pine Forge, 1733 (1733-1750), pp. 21, 166; New Pine Forge Ledger B, 1762-1763, p. 113. For an excellent study of colonial prices, see A Bezanson, R. D. Gray and M. Hussey, *Prices in Colonial Pennsylvania*, Chap. VI.

56 S. G. Hermelin, *Report About the Mines in the United States of America, 1783*, p. 621. It is hard to believe Hermelin's estimate that from forty-five tons of ore, thirty tons of pig iron could be produced. About two tons of ore usually made one ton of pig iron.

57 A. Bezanson, R. D. Gray and M. Hussey, *Prices in Colonial Pennsylvania*, Chap. VI.

58 S. G. Hermelin, *Report About the Mines in the United States of America, 1783*, p. 67.

59 J. D Schoepf, *Travels in the Confederation*, I, 199.

60 "Prices Current" in *Columbian Magazine*, July, 1789; *Universal Asylum and Columbian Magazine*, January, 1791, II, 63.

61 Israel Acrelius, *History of New Sweden*, p. 164.

62 Pine, Pool, Glasgow, Coventry, Helmstead, Windsor, Union and Pottsgrove Forges were reported. See M. A. Hanna, *Trade of the Delaware District Before the Revolution*, p. 258.

63 Tulpehocken Eisenhammer Account Books, 1754-1757; Durham Ledgers, Account Book of William Bird, pp. 105 ff. *passim*; Extracts and Memoranda of the Private Papers of John Taylor, Folio A, Cope Collection, p. 140.

64 See *Appendix* A for a list of forges.

65 *Pennsylvania Archives*, First Series III, 273.

66 The furnace ledgers show records of small amounts shipped to England and the West Indies during the period before 1750. Potts MSS. Colebrookdale Ledgers.

67 Israel Acrelius, *History of New Sweden*, p. 169.

68 4 George III, c. 15.

69 Israel Acrelius, *History of New Sweden*, p. 165.

70 *Pennsylvania Chronicle*, March 16-23, 1767; Cf. Harry Scrivenor, *History of the Iron Trade*, Appendix, p. 343.

71 *Pennsylvania Chronicle*, March 16-23, 1767.

72 *Pennsylvania Magazine*, February, 1775, pp. 72-73.

73 *Ibid.*, February, 1775, pp. 72-73; Cf. Harry Scrivenor, *History of the Iron Trade*, Appendix, p. 343-344.

74 *Pennsylvania Magazine*, February, 1775, pp. 72-73.

75 "Achenwall's Observations on North America, 1767," *Pennsylvania Magazine of History and Biography*, XXVII, 14.

76 Tench Coxe, *View of the United States*, p. 57 ff.

77 *Votes of the House of Representatives*, 1770, VI, pp. 200, 228. *Pennsylvania Chronicle*, October 21, 1772. *Pennsylvania Packet*, October 26, 1772; *Pennsylvania Packet Supplement*, January 18, 1773.

78 *American Archives*, Fourth Series, I, 1170-1171.

79 *Journals of the Continental Congress*, X, 56, 65.

80 Furnace MSS. Public Archives: *Pennsylvania Colonial Records*, XV, 229; *Pennsylvania Archives*, First Series, IV, 721, 735, 738, V, 31; Second Series, I, 487; *Pennsylvania Colonial Records*, X, 412, 530; *Pennsylvania Archives*, First Series, V, 36, 62, 64; *Pennsylvania Colonial Records*, X, 297; *Ibid.* X, 297. *Pennsylvania Archives*, Second Series, I, 485; Berkshire Furnace Ledgers.

81 *Pennsylvania Colonial Records*, XI, 335.

FOOTNOTES—Concluded

[82] J. D. Schoepf, *Travels in the Confederation*, I, 215.

[83] *Pennsylvania Archives*, First Series, IV, 731-732, V, 369; Second Series, I, 508, 574-576, Fourth Series, III, 610; *Pennsylvania Colonial Records*, X, 627, XI, 5, 6, 7, 20.

[84] *Ibid.*, X, 462, 500, 641; *Pennsylvania Archives*, First Series, VIII, 456.

[85] *Ibid.*, Second Series, I, 485; *Pennsylvania Colonial Records*, X, 533, 641, 648.

[86] *Ibid.*, X, 506; *Pennsylvania Archives*, First Series, IV, 712.

[87] John Penn to James Young, January 28, 1764, J. D. Rupp, *History of Dauphin and Cumberland Counties*, p. 401.

[88] J. D. Schoepf, *Travels in the Confederation*, I, 215.

[89] *Pennsylvania Archives*, First Series, IV, 708, 712, 717, 777; VI, 453, 475, 633; *Pennsylvania Colonial Records*, X, 290.

[90] *Pennsylvania Evening Post*, September 12, 1776, April 24, 1777.

[91] Potts MSS. LXX, Pine Forge, 1774 (1744-1781) pp. 78, 109, 130, 170, 188, 249, *Ibid.* LXXIII, Pine Forge, 1781 (1781-1782), pp. 10, 18.

[92] Tench Coxe, *View of the United States.* p. 490.

[93] *Ibid.*, p. 144.

[94] C. B. Montgomery Collection of MSS; *Philadelphia Federal Gazette*, February 13, 1799.

[95] *Freeman's Journal*, October 29, 1783, May 26, 1784, *Independent Gazetteer*, August 14, 1784, November 27, 1784, *Pennsylvania Gazette*, August 5, 1789, *Pennsylvania Packet*, June 22, 1795, *Federal Gazette*, Jan. 21, 1795, *Pennsylvania Packet*, September 4, 1797, *Philadelphia Federal Gazette*, July 31, 1798.

[96] *Philadelphia Federal Gazette*, November 29, 1793, April 21, 1795, September 1, 1796.

[97] John Oldmixon, *British Empire in America*, I xxiii, 318.

[98] Israel Acrelius, *History of New Sweden*, p. 143.

[99] *Cazenove Journal*, pp. 37-38.

[100] Tench Coxe, *View of the United States*, pp. 71-72.

[101] C. D. Ebeling, *Die Vereinter Staaten von Nordamerika*, IV, 89; Israel Acrelius, *History of New Sweden*, p. 167.

[102] C. D. Ebeling, *Die Vereinter Staaten von Nordamerika*, IV, 404.

[103] Tench Coxe, *View of the United States*, p. 313; see letter showing extent of manufactures of Lancaster in *Pennsylvania Gazette*, June 14, 1770, also J. D. Schoepf, *Travels in the Confederation*, II, 10 ff.; Lancaster County Historical Society Papers, II, 240.

[104] *Ibid.*, II, 20.

[105] Tench Coxe, *View of the United States*, p. 311.

[106] *Ibid.*, p. 415.

[107] *American Museum*, (1789) VI, *Ibid.* (1790) VII, 287.

[108] Tench Coxe, *View of the United States*, pp. 145-146

[109] *American State Papers: Finance*, II, 695.

APPENDIX A

Pennsylvania Ironworks of the Eighteenth Century[1]

Date of Origin	Name	Location	County	Erected by
1716	Rutter's Forge Works	Manatawny Creek	Berks	Thomas Rutter
1718	Coventry Iron Works	French Creek	Chester	Samuel Nutt, Sr.
About 1720	Colebrookdale Furnace	Iron Stone Creek	Berks	Thomas Rutter and Company
1722	Ball Bloomery	White Clay Creek	Newcastle (Del.)	John Ball
1725	Keith Furnace	Christiana River	Newcastle (Del.)	William Keith
1725	Pool Forge	Manatawny Creek	Berks	Alexander Wooddrop and Company
1725	Pine Forge	Manatawny Creek	Berks	Thomas Rutter
1725	McCall's (Glasgow) Forge	Manatawny Creek	Montgomery	George McCall
1726	Rock Run Furnace	French Creek	Chester	Samuel Nutt, Sr. and William Branson
About 1726	Kurtz (Bloomery) Forge	Octoraro Creek	Lancaster	Kurtz
1726	Abbington Furnace	Christiana River	Newcastle (Del.)	S. James and Company
1727	Durham Iron Works	Delaware River	Bucks	Anthony Morris and Company
1729	Spring Forge	Manatawny Creek	Berks	Anthony Morris
1732	Coventry Steel Furnace	French Creek	Chester	Samuel Nutt, Sr.
1733	Green Lane Forge	Perkiomen Creek	Montgomery	Thomas Maybury
Before 1733	Reading (Redding) Furnace No. 1	French Creek	Chester	Samuel Nutt, Sr. and William Branson
1736	Reading Furnace No. 2	French Creek	Chester	Samuel Nutt, Sr. and William Branson
1737	Warwick Furnace	French Creek	Chester	Anna Nutt and Company
1737	Peter Grubb Bloomery	Furnace Creek	Lebanon	Peter Grubb
1737	Mount Pleasant Furnace	Perkiomen Creek	Berks	Thomas Potts, Jr. and Co.
1739	Sarum Forge	Chester Creek	Delaware	John Taylor
1740	Crum Creek Forge	Crum Creek	Delaware	John Crosby and Peter Dicks
1742	Hopewell Forge	Hammer Creek	Lebanon	Peter Grubb
1742	Cornwall Furnace	Furnace Creek	Lebanon	Peter Grubb
1743	Mount Joy (Valley Forge)	Valley Creek	Chester	Daniel Walker and Company
1743	Mount Pleasant Forge	Perkiomen Creek	Berks	Thomas Potts, Jr.
1743	Windsor Forges	Conestoga Creek	Lancaster	William Branson

187

Date of Origin	Name	Location	County	Erected by
1744	Oley (Spang) Forge	Manatawny Creek	Berks	John Ross and Company
1744	Hopewell Forge	Union Township	Berks	William Bird
1744	New Pine Forges	Hay Creek	Berks	William Bird
1744	Darby Plating Mill	Darby Creek	Delaware	Solomon Humphreys
1745	Hereford Furnace	Perkiomen Creek	Berks	Thomas Maybury
1746	Sarum Slitting Mill	Chester Creek	Delaware	John Taylor
1746	Byberry Plating Mill	Byberry Township	Philadelphia	John Hall
1747	Tulpehocken Eisenhammer (Charming Forge)	Tulpehocken Creek	Berks	J. G. Nikoll and Company
About 1747	Paschall's Steel Furnace	Philadelphia	Philadelphia	Stephen Paschall
1747	Branson's Steel Furnace	Philadelphia	Philadelphia	William Branson
1750	Union Forge	Blue Mountains	Berks	
1750	Elizabeth Furnace	Middle Creek	Lancaster	John Jacob Huber
1750	Spring Forge (No. 2)	Pine Creek	Berks	
1750	Boiling Springs Forge	Boiling Springs	Cumberland	
About 1750	Quittapahilla (Newmarket) Forge	Quittapahilla Creek	Lebanon	Gerrard Etter
About 1752	Maria Forge	Poco Creek	Carbon	G. M. Weiss
1752	Pottsgrove Forge	Manatawny Creek	Berks	John Potts
1754	Offley's Anchor Forge	Philadelphia	Philadelphia	Daniel Offley
1754	Martic Furnace	Pequea Creek	Lancaster	T. and W. Smith
1755	Martic Forge	Pequea Creek	Lancaster	T. and W. Smith
1755	Roxborough (Berkshire) Furnace	Heidelberg Township	Berks	William Bird
1756	Dicks' Bloomery	Codorus Creek	York	Peter Dicks
Before 1759	Helmstead Forge	Furnace Creek	Berks	Benedict Swope and Company
1760	Shearwell (Oley) Furnace	Hammer Creek	Berks	James Old and D. Caldwell
1760	Speedwell Forge	Brandywine Creek	Lebanon	Mordecai Peirsol
About 1760	Rebecca Furnace	Maiden Creek	Chester	J Shöffer
1760	Moselem Forge	Furnace Creek	Berks	George Ross and Company
1762	Mary Ann Furnace	Boiling Springs	York	J. S. Rigby and Company
1762	Carlisle Furnace	French Creek	Cumberland	Thomas Potts and Company
1762	Vincent Forge and Steel Furnace	Pottsgrove	Chester	Samuel Potts and Company
1762	Pottsgrove Steel and Air Furnaces	Philadelphia	Montgomery	Whitehead Humphreys
1762	Humphreys' Steel Furnace	Mountain Creek	Philadelphia	R. Thornburg and Company
1764	Pine Grove Furnace	Codorus Creek	Cumberland	William Bennett
1765	Codorus Furnace and Forge		York	

Date of Origin	Name	Location	County	Erected by
Before 1766	Solebury Forge	Coryell's Ferry	Bucks	Frederic Delaplank
About 1766	Springton Forge	Brandywine Creek	Chester	Robert Thornburg
About 1766	Windsor Forge	Windsor Township	Berks	William Potts
1767	Thornburg Forge	Near Carlisle	Cumberland	Valentine Eckert
1768	District Furnace	Pine Creek	Berks	
1768	Windsor Furnace	Windsor Township	Berks	
1768	Gulf Forge	Gulf Creek	Montgomery	Mark Bird
1770	Hopewell Furnace	Union Township	Berks	George Ross and Company
1770	Spring Forge (No. 3)	Pigeon Hills	York	George Stevenson
1770	Mount Holly Iron Works	Mount Holly Springs	Cumberland	Mark Bird
1770	Gibraltar (Seyfert) Forge	Allegheny Creek	Berks	
1771	Salford Forge		Montgomery	Joseph and Samuel Potts
1775	Pine Forge Slitting Mill	Manatawny Creek	Berks	George Ege
1775	Charming Forge Slitting Mill	Tulpehocken Creek	Berks	Valentine Eckert and Henry Voight
Before 1776	Eckert and Voight's Wire Mill	Cumru Township	Berks	M. F. Alden
1778	Nanticoke Bloomery	Nanticoke Creek	Luzerne	
Before 1779	Birdsboro Slitting Mill and Steel Furnace	Birdsboro	Berks	Mark Bird
1779	Poole Forge	Caernarvon Township	Lancaster	James Old
1780	Oley Forges	Furnace Creek	Berks	Daniel Udree
1782	Carlisle Rolling and Slitting Mill and Steel Furnace	Near Carlisle	Cumberland	
1783	Rockland Forges	Beaver Creek	Berks	J. Truckenmiller
1783	Delaware Works (Forge and Slitting Mill)	Falls of Delaware River	Berks	Mark Bird and James Wilson
Before 1783	Mount Pleasant Iron Works	Path Valley	Franklin	W. Chambers and Brothers
1784	Twaddell's Forge	Brandywine Creek	Chester	William Twaddell
1784	Rockdale Forge	Ashton Township	Delaware	Abraham Pennell
1785	Bedford or Cromwell Furnace	Cromwell Township	Huntingdon	George Ashman, Thomas Cromwell, Charles Ridgely, and Tempest Tucker
Before 1785	Great Valley Works	Valley Creek	Chester	Isaac Potts and Company
About 1785	Buckley's (Brooke) Forge	Pequea Creek	Lancaster	Daniel Buckley
1785	Mount Hope Furnace	Chiquisalunga Creek	Lancaster	P. Grubb, Jr., and Company

189

Date of Origin	Name	Location	County	Erected by
About 1785	Mary Ann (Dowlin) Forge	Brandywine Creek	Chester	Dowlin
1785	Reading Forge	French Creek	Chester	Samuel Van Leer
Before 1786	Doe Run Forge	Doe Run	Chester	
1787	Thornbury Forge		Chester	Nancarrow and Matlack
1787	Nancarrow and Matlack's Steel Furnace	Philadelphia	Philadelphia	Nancarrow and Matlack
1789	Lackawanna Bloomery	Old Forge	Lackawanna	W. H. Smith
1789	Alliance (Jacob's Creek) Furnace	Jacob's Creek	Fayette	W. Turnbull and Company
1790	Union Furnace and Forge	Dunbar Township	Fayette	Isaac Meason
1790	Liberty Forge	Yellow Breeches Creek	Cumberland	
1790	Loudon Iron Works	Near Fort Loudon	Franklin	J. Chambers
1790	Pine Grove Forge	Pine Grove Run	Fayette	T. Lewis
1790	French Creek Forge	French Creek	Chester	C. Rentgen
1790	Phoenixville Slitting Mill and Nail Works	Phoenixville	Chester	B. Longstreth
1791	Lewis Forge	Robeson Township	Berks	Wm. Lewis
1791	Centre Furnace	Spring Creek	Centre	John Patton and Samuel Miles
1791	Colebrook (Mount Joy) Furnace	Conewago Creek	Lebanon	Robert Coleman
1791	Dale Furnace	Perkiomen Creek	Berks	Samuel Potts
1791	Licking Creek Forge	Licking Creek	Berks	T. Beale and Company
1791	Bedford Forge	Black Log Creek	Juniata	
1791	Sally Ann Furnace	Sacony Creek	Huntingdon	Valentine Eckert
1792	Antietem (Burkhart's) Forges	Antietem Creek	Berks	Samuel Potts
1792	Joanna Furnace	Hay Creek	Berks	S. Potts and Company
1792	Reading Furnace	Spring Creek	Berks	George Ege
1792	Hibernia Forge	Spring Creek	Berks	Samuel Van Leer
1792	Anshutz Furnace	West Brandywine Creek	Chester	G. Anshutz
1792	Hayden's Bloomery	Pittsburgh	Allegheny	John Hayden and John Nicholson
		George's Creek	Fayette	
1792	Reading Furnace	Spring Creek	Berks	George Ege
1793	Spring Grove Forge	Conestoga Creek	Lancaster	Cyrus Jacobs
1793	Rock Forge and Slitting Mill	Spring Creek	Centre	Philip Benner
1793	District Forge No. 1	Pine Creek	Berks	John Lesher
1793	Vincent Slitting Mill	French Creek	Chester	Potts and Hobart

Date of Origin	Name	Location	County	Erected by
1793	Hibernia Forge	Brandywine Creek	Chester	Caleb Foulke
1794	Foulke's Steel Furnace	Philadelphia	Philadelphia	B. & R. Jones
1794	Spring Hill Furnace	Ruble's Run	Fayette	Jeremiah Pears
1794	Plumsock Forge	Menallan Township	Fayette	G. P. Dorsey
1794	Barree Forge	Juniata River	Huntingdon	Michael Ege
1794	Cumberland Furnace	Yellow Breeches Creek	Cumberland	Jacob Lesher
1795	Mary Ann Furnace	Longswamp Township	Berks	William Brown and William Maclay
1795	Freedom Forge	Kishacoquillas Creek	Mifflin	Isaac Meason
1795	Mount Vernon Furnace	Mount's Creek	Fayette	Daniel Turner
1795	Spring Creek Forge	Spring Creek	Centre	Jacob Lesher
1795	District Forge No. 2	Pine Creek	Berks	Martin Dubbs and Company
1795	Westmoreland Furnace and Forge	Ligonier Valley	Westmoreland	Isaac Pennock
1795	Rokeby (Federal) Slitting Mill	Buck Run	Chester	Daniel Turner
1795	Turner's (Billington) Iron Works	Spring Creek	Centre	Miles, Dunlop and Company
1795	Harmony Forge (Milesburg Iron Works)	Spring Creek	Centre	
1795	Bellefonte Forge	Spring Creek	Centre	John Dunlop
1796	Sylvan Forge	George's Creek	Fayette	J. Oliphant
1796	Greenwood Furnace	Schuylkill Gap	Schuylkill	L. Reese and Company
1796	Huntingdon Furnace	Spruce Creek	Huntingdon	Mordecai Massey and Co.
1796	Schuylkill Forge	Schuylkill River	Schuylkill	George Ege
1797	Redstone Furnace	Redstone Creek	Fayette	Jeremiah Pears
1797	Laurel Furnace	Laurel Run	Fayette	J. Gibson and Company
1797	Hampton Forge	Laurel Run	Fayette	Reuben Mochabee
1797	Hope Furnace	Brightsfield	Mifflin	William Lewis
1797	Hall Forge and Tilt Mill	Spring Creek	Centre	John Hall
1797	Fairfield (Fairchance) Furnace and Forge	George's Township	Fayette	John Hayden
1798	Juniata Forge	Logan Township	Huntingdon	P. Shoenberger
1798	Spruce Creek Forge	Spruce Creek	Huntingdon	P. Massey
1798	Soundwell Forge	Roxbury	Franklin	Leephar, Crotzer and Company
1798	Logan Iron Works	Spring Creek	Centre	John Dunlop

Date of Origin	Name	Location	County	Erected by
1799	Germantown Steel Furnace	Germantown	Philadelphia	E. and B. Slocum
1799	Slocum's Hollow Bloomery	Roaring Brook	Lackawanna	William Lane
1800	Hopewell Forge		Bedford	P. Seidel
1800	Speedwell Forge	Angelica Creek	Berks	J. Webb
1800	Pine Grove Forge	Octoraro Creek	Lancaster	
1800	Sadsbury Forge	Octoraro Creek	Lancaster	J. Martin
1800	Mary Ann Furnace	Haydentown	Fayette	
1800	Little Falls (Bloomery) Forge	Arnold's Run	Fayette	Jeremiah Pears
1800	Pears' Slitting Mill	Menallen Township	Fayette	M. Withers
1800	Duquesne Forge	Octoraro Creek	Lancaster	Philip Benner
1800	Spring Furnace	Spring Creek	Centre	William McDermett
Before 1800	Caledonia Forge & Steel Furnace	Bedford Springs	Bedford	
Before 1800	Greene Furnace		Greene	
Before 1800	Youghiogheny (Lamb's) Forge	Youghiogheny River	Fayette	
Before 1800	Cheltenham Rolling and Slitting Mill		Montgomery	
Before 1800	Union Forges	Pine Creek	Berks	John Brobst

1 The few early ironworks of Delaware are also included.
This list is based largely upon information obtained from assessment lists, land office records, deeds, letters and other manuscripts. Locations in present counties are given.

APPENDIX B
Height and Size of Eighteenth Century Furnaces

Name of Furnace	County	Height	Size of Bosh
Pine Grove	Cumberland	33 feet	8½ feet
Huntingdon	Huntingdon	32 "	8 "
Union	Fayette	32 "	9 "
Mary Ann	Berks	30 "	7 "
Oley	Berks	30 "	9 "
Sally Ann	Berks	32 "	8 "
Joanna	Berks	28 "	7½ "
Hopewell	Berks	35 "	6½ "
Warwick	Chester	30 "	7½ "
Mount Hope	Lancaster	27 "	7 "
Colebrook	Lebanon	30 "	9 "
Cornwall	Lebanon	30 "	9 "
Centre	Centre	35 "	8 "
Fairchance	Fayette	35 "	9 "
Red Stone	Fayette	30 "	8 "

Alden's Appeal Record, Coleman *vs.* Brooke, p. 269.

APPENDIX C
Statement of Iron made at Cornwall Furnace

Year			Days out of Blust	Pig Iron and Castings
April 1, 1766 to March 31, 1767				945 tons 14 cwt.
"	1790	" 1791	134	595 " 10 "
"	1767	" 1768	86	862 " 11 "
"	1768	" 1769	72	750 " 14 "
"	1769	" 1770	30	908 " 15 "
"	1770	" 1771	90	861 " 5 "
"	1771	" 1772		691 " 1 "
"	1772	" 1773		
"	1773	" 1774	83	744 " 2 "
"	1774	" 1775		
"	1775	" 1776		
"	1776	" 1777		353 "
"	1777	" 1778		488 " 8 "
"	1778	" 1779		183 "
"	1779	" 1780		70 "
"	1780	" 1781		263 " 7 "
"	1781	" 1782		186 " 15 "
"	1782	" 1783		361 " 11 "
"	1783	" 1784		434 " 15 "
"	1784	" 1785		285 " 12 "
"	1785	" 1786		483 " 10 "
"	1786	" 1787	175	468 " 19 "
"	1787	" 1788	147	507 " 11 "
"	1788	" 1789	188	650 "

APPENDIX C—Continued

Year	Days out of Blast	Pig Iron and Castings
April 1, 1789 to March 31, 1790	73	560 tons
" 1790 " 1791	134	595 " 10 cwt.
" 1791 " 1792	80	643 "
" 1792 " 1793		724 "
" 1793 " 1794		844 "
" 1794 " 1795		845 " 12 "
" 1795 " 1796		865 "
" 1796 " 1797		930 "
" 1797 " 1798		796 "
" 1798 " 1799	45	1,347 "
" 1799 " 1800	34	1,457 " 18 "

Alden's Appeal Record, Coleman *vs.* Brooke, p. 157.

APPENDIX D

Statement of Iron made at Elizabeth Furnace

Year	Pig iron and Castings	Length of Blast	Average per week
1780	269 tons 5 cwt.	20 weeks	13.5 tons
1781	358 " 15 "	23 "	15.5 "
1782	413 " 15 "	26 "	16 "
1783	526 " 17 "	28 "	19 "
1784	461 " 17 "	23 "	20 "
1785	558 " 2 "	22 "	25 "
1786	459 " 2 "	15 "	30 "
1787	664 "	22 "	30 "
1788	651 " 15 "	31 "	21 "
1789	717 " 7 "	32 "	23 "
1790	753 " 16 "	30 "	25 "

Alden's Appeal Record, Coleman *vs.* Brooke, p. 134.

APPENDIX E

Statement of Iron made at Colebrook Furnace

Year	Pig Iron	Castings	Total
1797	1,069 tons	75 tons	1,144 tons
1798	1,200 "	63 "	1,263 "
1799	1,317 "	68 "	1,385 "
1800	1,466 "	47 "	1,513 "

Alden's Appeal Record, Coleman *vs.* Brooke, p. 134.

APPENDIX F[1]

Articles of Copartnership indented made & agreed upon this Ninth Day of June in the Year of our Lord One thousand seven hundred and seventy two By and between Mark Bird Esqr. William Dewees junr. and Joseph Potts all of the County of Philadelphia Iron masters in manner following: that is to say Imprimis That the said Parties from the Day of the Date hereof for and during the Space of six consecutive Years shall and will be jointly and eqally concerned in the making or Manufacturing of Barr Iron at the Forge called Glasgow which with the Lands thereto belonging they have purchased of Archibald McCall. Item that the said William Dewees and Joseph Potts shall and will pay for & advance all the Stock alive and Dead necessary for carrying on the said Forge Works except the Pigg Iron and for manageing the Plantations or farms on the said Lands. And the said Mark Bird shall be charged in the Account of the said partnership with so much as each of the said William Dewees & Joseph Potts shall have advanced towards the cost of such Stock to be rendered or made up by him the said Mark Bird in Pigg Iron at the rate herein after mentioned Item that the said Mark Bird shall and will from time to time and at Times during the Continuance of the said Partnership from and out of his Furnace called Hopewell whilst the same is in Condition to be worked or from any other place not exceeding twelve Miles distance from Glasgow Forge (Colebrook Dale all other Coale share Metals excepted) furnish the Company of Glasgow Forge with as many Piggs as the same Forge is able to work into Barr Iron to be dilivered when called for at his said Hopewell Furnace Banck or at such other place as aforesaid charging the Company for every Ton of his said Pigg Iron with one fourth part of the current Market price of Barr Iron in the City of Philadelphia and two Shillings and sixpence over and above such fourth part And it is agreed that by the delivery of the said Pigg Iron at the rate aforesaid the Moiety of the Stock which shall have been advanced by them the said William Dewees & Joseph Potts as aforesaid shall be first made up or ballanced and from thenceforth for all future Deliveries of Pigg Iron the said Mark Bird shall be paid at the rate aforesaid out of the Moneys belonging to the said Company whenever his account of Piggs delivered shall amount to Twenty five Ton — Item That when there is no Cash belonging to the said Company in the hands of either of them the said Partners all the wages and dieting of hired Men and Servants the Repair of the Works and other incidental Charges of carrying on the same shall be immediately as it is wanted paid & disbursed by all the said parties equally. Item that all the Barr Iron which shall not be sold or contracted for at the Forge for Country use or bartered for Goods & chattled to the use of the Company shall be sent to the store or Stores of the said Mark Bird in the city of Philadelphia whilst he shall be residing here to be sold by him for the Benefit of the Partnership and for his trouble he shall be allowed the sum of Twenty Shillings per Ton Commissions on the whole sale thereof. Item that the said Joseph Potts whist he shall reside at Pottsgrove shall attend to and direct the Management of the said Forge and Plantations and oversee the works of the Persons employed therein and especially take Care of the Books of Accounts & Vouchers of the said Partnership so that the same may be kept just and clear and always ready for the access and Inspection of the said Partners. And for his Trouble in the Premises the said Joseph Potts shall be allowed at the rate of One hundred and thirty Pounds pr Ann he paying wages for a Clerk and the Company finding his Diet and provender for his Horse. Item that at the Decease either of the said Partners, within The Space of six Months next after the Accounts & State of their affairs shall be made up and settled between the Survivors and the Executors of Administrators of such Deceased and his interest as well in the Lands Tenements and Hereditiments purchased of the said Archibald McCall as in the Stock, & Effects, Goods, & Chattles belonging to the said Partners shall be valued by three or more Persons to be chosen and appointed by the mutual Consent of the Survivors and Executors or Administrators of such Deceased Partner And the Survivors shall have right to take the part of such Deceased at the valuation so to be made as aforesaid or on their refusal the Heirs Executors or Aministrators of such deceased Partner may either continue his Interest for the benefit of his Estate or sell the same as they think fit and the said Mark Bird for himself his Executors and Administrators doth Covenant promise grant and agree to and with the said William Dewees and Joseph Potts their several and respective Executors and Administrators that the said Mark Bird his Exors & Administrators shall and will well and truly observe do and perform all and singular the Matter and Things which by virtue of the above agreement and according to the true meaning and Intent thereof on his or their part are or ought to be Observed done and performed so that the Works of the said Partners and carrying on their joint Trade and Business shall or may not be obstructed or delayed for or on account of his or their Neglect non repliance or Refusal And the said William Dewees for himself his Executors and Administrators doth Covenant promise and agree and Grant to and with the said Mark Bird and Joseph Potts their several and respective Executors and Administrators that the said William Dewees his Executors and Administrators Shall and will well and truly observe do And performe all & Singular the Matters and things which by virtue of the above agreement & according to the true meaning and Intent thereof on his or their Part are or ought to observe done or perform so that the Works of the said Partners and carrying on their Business and joint Trade shall or may not be Obstructed or delayed for or on Account of his or their Neglect non Compliance or refusal. And the said Joseph Potts himself his Executors & Administrators doth Covenant promise Grant and agree to and with the said Mark Bird and William Dewees their several and respective Executors and Administrators that the said Joseph Potts his Executors & administrators shall and will well & truly observe do and perform all and singular the Matters and Things which by virtue of the above agreement and according to the true Meaning and Intent thereof on his or their part are or ought to be observe done and performed So that the Works of the said Partners and carrying on their joint Trade and Business shall or may not be obstructed or delayed for or on Account of his or their non Compliance or refusal. In Witness Whereof the said parties have hereunto set their hands and Seals Dated the Day and Year first above written.

[1] Because of the financial difficulties connected with operating ironworks, agreements similar to the above were common during the eighteenth century.

BIBLIOGRAPHY

Manuscript Sources

American Philosophical Society, MSS. Communications to:
Mechanics, Machinery, and Engineering. Library, American Philosophical Society.

American Philosophical Society, MSS. Communications to:
Navigation, Manufactures, Agriculture and Economics.
Library, American Philosophical Society.

Bancroft Collection: Colonial Documents, 1748-1774. Letters and documents of the Board of Trade, colonial governors, and others, on American affairs. 535 Transcripts. 7 vols. New York Public Library.

Bancroft Collection: Massachusetts Papers, 1748-1755. New York Public Library.

Bancroft Collection: Thomas Penn's Letters, 1748-1755. New York Public Library.

Bartram Papers. Vol. II. Historical Society of Pennsylvania.

Berks and Montgomery County MSS., 1693-1869. Historical Society of Pennsylvania.

Berkshire (Roxborough) Furnace Account Books, 1767-1781. Ledgers, journals, day books and waste books. Historical Society of Pennsylvania.

Berkshire Furnace MSS., 1777-1793. Private Collection.

Bird William. Account Books, 1741-1765. Ledgers and receipt book pertaining to iron manufacture in Berks County. Historical Society of Pennsylvania.

Birdsborough (Birdsboro) Iron Works MSS.: Correspondence and Account Book, 1788. Historical Society of Berks County.

Board of Trade Journals: 1675-1782. Transcribed from the original manuscripts in the British Public Record Office. 93 vols. Historical Society of Pennsylvania. A part of this work has been printed under the title of *Journal of the Commissioners for Trade and Plantations* (1704-1767).

Board of Trade Papers: Plantations General, 1689-1780. Transcribed from the original manuscripts in the British Public Record Office. 31 vols. Historical Society of Pennsylvania.

Board of Trade Papers: Proprieties, 1697-1776. Transcribed from the original manuscripts in the British Public Record Office. 24 vols. Historical Society of Pennsylvania.

Chalmers Collection. Original documents and transcripts relating to America, collected by George Chalmers. 25 vols. New York Public Library.

Charming Forge MSS.: Account books of Charming Forge, 1790-1832. Historical Society of Berks County.

Charming Forge, MSS., 1784-1800. Private Collection.

Clifford Papers, 1722-1832. 29 vols. Historical Society of Pennsylvania.

Colebrook MSS., 1792-1800. Private Collection.

Colebrookdale Furnace Account Books, 1750-1766. (Potts MSS.) A series of ledgers and day books. Historical Society of Pennsylvania.

Colebrookdale (Popodickon). Furnace Account Books, 1744-1751. (Potts MSS.) Ledgers and day books of the first blast furnace built in Pennsylvania, 1720. Historical Society of Pennsylvania.

Coleman, Robert, MSS., 1783-1800. Private Collection.

Cope, Gilbert. Collections: Papers Relating to Manufactures Pertaining Chiefly to the Iron Industry in Chester County. Historical Society of Pennsylvania.

Cornwall Furnace MSS., 1764-1800. Private Collection.

County Archives. Deed books, wills, road dockets, township property rolls, assessment lists etc. Philadelphia, Bucks, Chester, Berks, Lancaster, York, Cumberland, Mifflin, Huntingdon, Westmoreland, Fayette and Allegheny Counties.

Coventry Forge MSS.: Account Book, 1789-1790. Chester County Historical Society.

Coventry Ironworks Account Books, 1726-1760 (Potts MSS). A series of ledgers and day books. Historical Society of Pennsylvania.

Craig MSS. Letter Book of Henry Knox, Secretary of War. Letter Books C. G. Invoice Book of Isaac Craig. Carnegie Library of Pittsburgh.

Cuthbertson, John. Diary. The observations of an eighteenth century Reformed Presbyterian minister. The diary contains information pertaining to social conditions and mentions several iron plantations. Pittsburgh Theological Seminary of the United Presbyterian Church.

Dale Furnace Account Books, 1799-1801 (Potts MSS). Ledger and Journal. Historical Society of Pennsylvania.

Dartmouth Papers. State of the British Islands in the West Indies. Vol. I. William L. Clements Library, Ann Arbor, Mich.

Dickinson, Jonathan. Letter Book, 1713-1721. Ridgeway Branch, Library Company of Philadelphia.

Drinker, Henry. Letter Books, 1773-1800. 11 vols. Contains letters giving valuable information regarding iron manufacture and prices. Historical Society of Pennsylvania.

Durham, Iron Works Records, 1778-1800. (Backhouse MSS.) Ledgers, account books and journals. Bucks County Historical Society.

Ege Collection MSS. Correspondence and memoranda about George Ege's interests in ironworks 1783-1784. Historical Society of Berks County.

Elizabeth Furnace Account Books, 1756-1779. Ledgers, day books and a journal of Manheim Glass Works. Historical Society of Pennsylvania.

Elizabeth Furnace MSS., 1768-1800. Private Collection.

Fackenthal, B. F., Jr. Briefs of Title to Real Estate in Durham and Adjoining Townships in Pennsylvania. Historical Society of Pennsylvania.

Franklin, Dr. Benjamin. Miscellaneous Papers and Letters of. American Philosophical Society.

Great Britain. Colonial Office Papers. Transcripts 5:338; 5:886; 5:889. Public Record Office.

Great Britain. Colonial Office Papers. Transcripts. 5:1281; 5:1319; 5:1320; 5:1323; 5:1324. Library of Congress.

Great Britain. House of Lords MSS. Transcripts. 86, 88, 125, 135, 183, 185. Library of Congress.

Great Britain. British Museum, King's MSS. Transcripts. 205, 206. Library of Congress.

Great Britain. British Museum, Additional MSS. Transcripts 29600, 33029, 35908, 35909. Library of Congress. For a description of the above British MSS., see C. M. Andrews, *Guide to the Materials for American History to 1783, in the Public Record Office of Great Britain.* 2 vols. Washington, 1912, 1914. Also, C. M. Andrews and F. G. Davenport, *Guide to the Manuscript Materials for the History of the United States to 1783, in the British Museum, in Minor London Archives, and in the Libraries of Oxford and Cambridge.* Washington, 1908.

Green Lane Forge Books. Private Collection.

Hardwicke MSS. Vol. DLXII. Library of Congress.

Hopewell Furnace Account Books, 1784-1808. Ledgers and day books. Historical Society of Pennsylvania.

James, Mrs. T. P. Manuscript Collections. Papers relating to the ironworks owned and managed by various members of the Potts family. Historical Society of Pennsylvania.

Krick, Anna. Manuscripts. Berks County Historical Society.

Linn, J. B. MSS. 1790-1800. Private Collection.

Logan Letter Book. (Parchment Book) Historical Society of Pennsylvania.

Logan Papers. Vol. XXXVIII. Historical Society of Pennsylvania.

Lynn, Mass. Iron Works MSS. 1650-1685. Baker Library, Business Historical Society.

Manheim Glass Works MSS. Historical Society of Pennsylvania.

Mary Ann Furnace Account Books, 1762-1776. Ledgers, day books, journals and waste books. Historical Society of Pennsylvania.

Mary Ann Furnace MSS., 1764-1800. Private Collection

May, Thomas. Account Books, 1762-1777. Books (Dreer Collection) Historical Society of Pennsylvania.

Mitchell, J. Thomas, MSS. Private Collection.

Mount Joy (Colebrook) Furnace MSS., 1791-1797. Private Collection.

Mount Pleasant Furnace Account Books, 1737-1747 (Potts MSS.). Ledgers and day books. Historical Society of Pennsylvania.

Montgomery, C. B. Collection of MSS. Documents, papers and account books relating to the early iron industry in the Schuylkill Valley, Pennsylvania. Private collection.

New Pine Forge (Birdsboro). Account Books, 1744-1778. Ledgers, day books, journals and charcoal books. Historical Society of Pennsylvania.

New Pine Forge MSS.: Account Books, 1762. Historical Society of Pennsylvania.

Penn Family, Letters of, to James Logan, 1701-1729. 2 vols. Historical Society of Pennsylvania.

Penn Letter Books. Vols. I-XIV. Historical Society of Pennsylvania.

Penn Manuscripts. Indian Affairs, 1757-1772. Historical Society of Pennsylvania.

Penn-Physick Papers. Vol. XV. Historical Society of Pennsylvania.

Penn MSS. Vols. I-XIV. Vol. XIII contains many valuable documents and pamphlets written and printed in England during the struggle over the regulation and control of the colonial iron industry. Historical Society of Pennsylvania.

Penn MSS. Charters and Frame of Government. Historical Society of Pennsylvania.

Penn Official Correspondence, 1683-1817. 12 vols. Historical Society of Pennsylvania.

Pennsylvania Furnace MSS. Archives of Pennsylvania, State Library, Harrisburg, Pa.

Pennsylvania Miscellaneous Collection, 1777. Library of Congress.

Pennsylvania Miscellaneous Papers: Penn and Baltimore, Penn Family, 1740-1756. Historical Society of Pennsylvania.

Pennsylvania, Proceedings of the Provincial Conference, June 18-June 25, 1776. Historical Society of Pennsylvania.

Pine Forge Account Books, 1720-1790 (Potts MSS). A series of ledgers and day books. Historical Society of Pennsylvania.

Pool Forge MSS, 1725-1853. Historical Society of Berks County.

Pool Forge MSS: Journal, 1749-1759. Historical Society of Pennsylvania.

Potts Miscellaneous Collection, 1738-1840. Chiefly account books pertaining to milling, the sale of merchandise, and surveying. Historical Society of Pennsylvania.

Potts Philadelphia Account Books, 1737-1776. Ledgers, day books and journals that record the commercial activities of the Potts family in Philadelphia. Historical Society of Pennsylvania.

Pottsgrove (Pottstown) MSS., 1755-1825. Ledgers and day books that record the commercial and milling activities of the Potts family. Pottstown which became an early industrial center was laid out in 1752 by John Potts, Sr. Historical Society of Pennsylvania.

Reading Furnace MSS., 1793-1800. Private Collection.

Sarum Forge Account Books, 1767-1771. Ledger and day book. Historical Society of Pennsylvania.

Schuylkill Forge MSS., 1796-1800. Private Collection.

Smith, Charles Morton. MSS. Historical Society of Pennsylvania.

Speedwell Forge MSS., 1784-1800. Private Collection.

Spring Forge (York County) MSS., 1765-1800. Private Collection.

Strettel MSS. Historical Society of Pennsylvania.

Tulpehocken Account Book, 1744-1749 (Potts MSS). Iron and groceries sold by John Potts, Sr. in Berks, Lebanon and Dauphin counties. Collections made by Conrad Weiser. Historical Society of Pennsylvania.

Tulpehocken Forge Account Books, 1754-1791. Ledgers, journals and day books. Historical Society of Pennsylvania.

Union Forge MSS., 1783-1800. Private Collection.

Valley Forge (Mount Joy) Account Books, 1757-1767 (Potts MSS). A series of ledgers, day books and journals. Historical Society of Pennsylvania.

Warwick Furnace Account Books, 1742-1773. (Potts MSS). A series of ledgers, day books, journals and receipt books. Historical Society of Pennsylvania.

Wayne MSS., Vol. IV. Correspondence and papers of Anthony Wayne. Historical Society of Pennsylvania.

Wilson, James. Papers of. Vols. III, VII. Historical Society of Pennsylvania.

Official Printed Sources

American Archives, Edited by Peter Force. Fourth Series, 6 vols. Fifth Series, 3 vols. Washington, 1837-1853.

American State Papers, Claims. Vol. 1. Washington, 1832.

American State Papers, Finance. Vol. 1. Washington, 1832.

American State Papers, Foreign Relations. Vols. I, II. Washington, 1832.

Great Britain, *Acts of the Privy Council of England. Colonial Series. Rolls. Series.* Edited by W. L. Grant and James Munro. Vols. I-VI (1613-1783) Hereford, 1908-1912.

Great Britain, *Calendar of State Papers, Colonial Series, American and West Indies.* Vols. I-XXX (1574-1718) London, 1860-1930.

Great Britain, *Calendar of State Papers. Colonial Series, East Indies, 1513-1616.* Vol. II. London, 1862.

Great Britain. *Debates and Proceedings of the House of Commons*, 1784-1785. 5 vols. London, 1785.

Great Britain, *Journals of the House of Commons*, 1640-1800. Vols. II-XLII. 1803.

Great Britain, *Journals of the House of Lords*, 1660-1800. Vols. II-XLII. 1817-1836.

Great Britain, *Parliamentary History of England.* Edited by William Cobbett. Vols. I-XXXIV. London, 1806-1819.

Great Britain, *Report of a Committee of the Lords of the Privy Council on the Trade of Great Britain with the United States*, 1791. Edited by W. C. Ford. Washington, 1888.

Great Britain, *Statutes at Large from Magna Charta to . . .* 1763. Compiled by Owen Ruffhead. 9 vols. London, 1763-1765. New edition revised and continued to 1800 by Charles Runnington. 14 vols. London, 1786-1800.

Great Britain, *The Statutes; Second Revised Edition* (1235-1886). 16 vols. London, 1883-1900.

Maryland, *Archives of*. Edited by W. H. Browne and others. Vols. I-XLVII. Baltimore, 1883-1930.

Maryland, *Laws of*. Compiled by Thomas Bacon. Annapolis, 1765.

Massachusetts, *Essex County, Records and Files of the Quarterly Courts*. Edited by J. F. Dow. Vols. I-VIII. Salem, 1911-1921.

Massachusetts, *Records of the Governor and Company of the Massachusetts Bay in New England* (1628-1686). Edited by N. B. Shurtleff. 5 vols. Boston, 1853-1854.

New Jersey, *Archives*. Edited by W. A. Whitehead and others. First and Second Series. 30 vols. Newark, etc., 1880-1906.

New York, *Documents Relative to the Colonial History of the State of New York*. Edited by E. B. O'Calloghan and B. Fernow. 15 vols. Albany, 1856-1887.

North Carolina, *Colonial Records of* 1662-1776. Edited by W. L. Saunders. 1886-1906.

Pennsylvania Archives, Seven series. 101 vols. Philadelphia and Harrisburg, 1852-1914.

Pennsylvania Colonial Records. Minutes of the Provincial Council and of the Supreme Executive Council. Edited by Samuel Hazard. 16 vols. Harrisburg, 1852-1853.

Pennsylvania, *Duke of Yorke's Book of Laws* (1676-1682). Harrisburg, 1879.

Pennsylvania, *Journals of the House of Representatives* (November 28, 1776 to October 2, 1781). Michael Hillegas, ed. Philadelphia, 1782.

Pennsylvania, *Journals of the House of Representatives*, 1790-1800. Philadelphia, 1790-1800.

Pennsylvania, *Minutes of the General Assembly*, 1781-1790. Philadelphia, 1790.

Pennsylvania, *Votes and Proceedings of the House of Representatives of the Province of Pennsylvania*. (1682-1776) 6 vols. Philadelphia, 1752-1776. Commonly called *Votes of the Assembly*.

Pennsylvania Statutes at Large, 1700-1809. Edited by James T. Mitchell and Henry Flanders. Vols. II-XVIII. Harrisburg, 1896-1915.

United States, Congress, *Annals of*. 1789-1824. 42 vols. Washington, 1834-1856.

United States, Congress—*House Documents*, 55th Cong. 2nd Sess. No. 562. Contains tariff laws from 1789. Washington, 1898.

United States, *Journals of Congress*, 1774-1788. 13 vols. Philadelphia, 1777-1789; reprint known as *Journals of the American Congress*. 4 vols. Washington, 1823.

United States, *Journals of the Continental Congress*. Edited from the original by W. C. Ford and Gaillard Hunt. Vol. I. 1774. Washington, 1904.

United States, *Treaties, Conventions, International Acts, etc.* Edited by W. M. Malloy. Vol. I. Washington, 1910.

United States, *Secret Journals of the Acts and Proceedings of Congress.* 4 vols. Boston, 1821.

United States, *Third Census,* 1810.

Virginia Company of London, *History and Letters to and from the First Colony.* Edited by Edward D. Neill. New York, 1869.

Virginia Company of London, *Records of the: Court Book, from the Manuscript in the Library of Congress.* Edited by S. M. Kingsbury. 2 vols. Washington, 1906.

Wharton, F., *Diplomatic Correspondence of the American Revolution.* Vol. VI. Washington, 1889.

General Printed Sources

Abbott, Benjamin, *Experience and Gospel Labors.* Philadelphia, 1825.

Acrelius, Israel, *History of New Sweden, or Settlements on the River Delaware.* Translated from the Swedish by W. M. Reynolds. Historical Society of Pennsylvania, *Memoirs* IX. Philadelphia, 1874.

Agricola, Georgius, *De Re Metallica.* Translated from the first Latin edition of 1556 by H. C. Hoover and L. H. Hoover. Published by Mining Magazine, London, 1912.

Almond, John, *A Collection of the Most Interesting Tracts on the Subjects of Taxing the American Colonies and Regulating their Trade.* 3 vols. London, 1766-1767.

American Husbandry. London, 1775.

Anburey, Thomas, *Travels through the Interior Parts of America,* 2 vols. London, 1791.

Anderson, Adam, *Historical and Chronological Deduction of the Origin of Commerce,* 4 vols. London, 1787-1789.

Ashe, Thomas, *Travels in America in 1806.* New York, 1811.

Baker, W. S. *Itinerary of George Washington.* Philadelphia, 1892.

Bayard, M., *Voyage dans l'Interieur des Etats Unis, 1791.* Paris, 1791.

Beaujour, Chevalier Felix de, *Apercu des Etats Unis, 1800-1810.* Paris, 1814.

Bernard, John, *Retrospections of America, 1797-1811.* Edited from the Manuscript, by Mrs. Bayle Bernard. New York, 1887.

Bonnet, J. E., *Etats-Unis de L'Amerique a la fin du XVIII Siecle.* 2 vols. Paris, 1802.

Bradbury, John, *Travels in the Interior of America, 1809, 1810 and 1811.* London, 1819.

Brissot, J. P., *The Commerce of America with Europe.* New York, 1795.

Brissot, J. P., *Noveau Voyage dans les Etats Unis de L'Amerique Septentrionale fait en 1788.* 3 vols. Paris, 1791.

Brown, Alexander, *Genesis of the United States.* 2 vols. Boston, 1890. A documentary history of Virginia to 1616.

Budd, Thomas, *Good Order Established in Pennsylvania and New Jersey,* 1685. New York, 1865.

Bulow, D. von, *Der Freistaat von Nordamerika, 1797.* Berlin, 1797.

Burnaby, Rev. Andrew, *Travels through the Middle States of North America, 1759-1760.* London, 1775.

Byrd, William, *The Writings of "Colonel William Byrd of Westover in Virginia, Esquire."* Edited by John Spencer Basset, New York, 1901.

Carver, Jonathan, *Three Years Travels through the Interior Parts of America.* Edinburgh, 1798.

Cazenove, T. *Journal, 1794.* Translated and edited by R. W. Kelsey. Haverford, 1922.

Coke, Thomas, *A Journal of the Rev. Dr. Coke's Fourth Tour on the Continent of America.* London, 1792.

Columbus, Christopher, *Journal of First Voyage to America.* Albert and Charles Boni, New York, 1924.

Columbus, Christopher, *Letters and Other Original Documents.* Publications of the Hakluyt Society, Series I, Vol. XLIII, London, 1870.

Cooper, Thomas, *Some Information Respecting America, 1793.* Dublin, 1794.

Coxe, Tench, *View of the United States.* Philadelphia, 1794.

Crevecoeur, M. G. J. de, *Voyage dans la Haute Pennsylvanie.* 3 vols. Paris, 1801.

Davis, John, *Travels of Four Years and a Half in the United States of America, 1798-1802.* New York, 1909.

Doddridge, Joseph, *Notes on the Settlement and Indian Wars of the Western Parts of Virginia and Pennsylvania.* Albany, 1876.

Douglass, William, *British Settlements in North America.* A Summary Historical and Political, 2 vols. Boston, 1749-1753.

Drinker, Elizabeth, *Extracts from the Journal of, 1759-1809.* Edited by Henry D. Biddle, Philadelphia, 1889.

Dudley, Dud, *Mettalum Martis,* London, 1665.

Ebeling, C. D. *Die Vereinten Staaten von Nordamerika, in Erdbeschreibung und Geschichte von Amerika.* Vols. IV and V. Hamburg, 1797, 1799.

Ecuyer, A. W. S. *Lettres d'un Americain Cultivateur.* Maestrichte, 1789.

Eden, Richard (Ed.), *The First Three English Books on America.* Westminster, 1865.

Ellicott, Andrew, *Journal, 1796-1800.* Philadelphia, 1803.

Encyclopedia Britannica, or *Dictionary of Art, Sciences, etc. London,* 1797.

Evans, Oliver, *Patent Right Oppression Exposed or Knavery Detected.* By Patrick N. I. Elisha. Philadelphia, 1813.

Fearon, H. B. *Narrative of a Journey through the Eastern and Western States.* London, 1818.

Force, Peter, *Tracts and Other Papers.* Dealing with the Settlement and Progress of the Colonies to 1776. 4 vols. Washington, 1836-1846.

Frame, Richard, *A Short Description of Pennsylvania, 1692.* In metrical verse. Reprinted from the unique copy in the Philadelphia Library. Philadelphia, 1867.

Franklin, Benjamin, *Autobiography.* Everyman's Library. New York, 1916.

Franklin, Benjamin, *A Modest Enquiry into the Nature and Necessity of a Paper Currency.* Philadelphia, 1729.

Franklin, Benjamin, *Works*. Edited by Albert H. Smyth. 10 vols. New York, 1905-1907.

Hamilton, Alexander, *Itinerarium, A Narrative of a Journey through Maryland, Delaware, Pennsylvania*. St. Louis, 1907. This journey was made in 1744.

Hamilton, Alexander, *Report of the Secretary of the Treasury on Manufactures*, 1791. Philadelphia, 1824.

Hariot, Thomas, *A Briefe and True Report of the New Found Land of Virginia, 1590*. Reprinted, London, 1843.

Hasenclever, Peter, *The Remarkable Case of*. London, 1773. Photographic copy in New York Public Library.

Hermelin, S. G. *Report About the Mines in the United States of America, 1783*. Translated from the Swedish by Amandus Johnson. Philadelphia, 1931.

Herrera, T. A. *Historia General*. 9 vols. in 5. Madrid, 1726-1736.

Hiltzheimer, Jacob, *Extracts from Diary, 1765-1798*. Edited by J. C. Parsons, Philadelphia, 1893.

Holm, Thomas Campanius, *A Short Description of the Province of New Sweden*. Translated by Peter S. DuPonceau. Philadelphia, 1834.

Horne, Henry, *Essays Concerning Iron and Steel*. London, 1773.

Humphries, David, *Poem on Industry*. Philadelphia, 1794.

Interests of the Merchants and Manufacturers of Great Britain in the Present Contest with the Colonies Stated and Considered. Philadelphia, 1774.

Jars, M. *Voyages Metallurgiques*, 3 vols. 1774.

Kalm, Peter, *Travels into North America*. Translated by John R. Forster. 2 vols. London, 1770, 1772.

Keith, William, *A Just and Plain Vindication*, Philadelphia, 1726.

La Rochefaucauld, Liancourt, F. A. F., *Voyage dans les Etats-Unis d'Amerique, fait en 1795-1797*. 8 vols. Paris, 1799.

Lasso de la Vega, Garcilasco. *Commentarios Reales*. 2 vols. Madrid, 1722-1733.

Latrobe, B. H. L. *Journal of Latrobe*. New York, 1905.

Letter to a Member of Parliament Concerning the Naval Store Bill. London, 1720.

Logan, James, *A More Just Vindication of the Honourable Sir William Keith*. Philadelphia, 1726.

Mascall, E. J. *Book of Customs*. London, 1801.

Matthews, W. *Historical Review of North America*. 2 vols. Dublin, 1789.

MacPherson, David, *Annals of Commerce*. 4 vols. London, 1805.

Melish, John, *Travels through the United States of America, 1806-1811*. 2 vols. Philadelphia, 1815.

Michaux, F. A., *Travels to the West of the Alleghany Mountains*. London, 1805.

Mitchell, John, *The Present State of Great Britain and North America with Regard to Agriculture, Population, Trade and Manufactures*. London, 1767.

Mittleberger, Gottlieb, *Journey to Pennsylvania in the Year 1750*. Translated by C. J. Eben. Philadelphia, 1898.

Morris, Lewis, *Papers of.* New Jersey Historical Society, Collections. Vol. IV. New York, 1852.

Oldmixon, John, *The British Empire in America.* 2 vols. London, 1741.

Oviedo y Valdes, Gonzalo Fernandez. *Historia General y Natural de Las Indias.* 4 vols. Madrid, 1851.

Parkinson, Richard, *A Tour in America, 1798, 1799, 1800.* 2 vols. London, 1805.

Penn-Logan Correspondence, 1700-1750. Edited by Edward Armstrong. Historical Society of Pennsylvania, *Memoirs.* Vols. IX and X. Philadelphia, 1870-1872.

Penn, William, *Some Account of the Province of Pennsylvania.* London, 1681.

Penn, William, *A Further Account of the Province of Pennsylvania and its Improvements.* London, 1685.

Priest, William, *Travels in the United States of America Commencing in the Year 1793 and Ending in 1797.* London, 1802.

Proud, Robert, *History of Pennsylvania.* 2 vols. Philadelphia, 1797-1798.

Reaumur, R. A. F. de, *L'Art de Convertir le Fer Forge en Acier.* Paris, 1722.

Rees' Encyclopedia. Philadelphia, 1802.

Schoepf, Johann David, *Travels in the Confederation, 1783-1794.* Translated by A. J. Morrison, 2 vols. Philadelphia, 1911.

Scott, Joseph, *Gazetteer of the United States. Philadelphia,* 1795.

Sheffield, John Lord, *Observations on the Commerce of the United States.* London, 1784.

Smith, Richard, *A Tour of Four Great Rivers in 1769.* New York, 1910.

Spotswood, A., *Official Letters of Governor, 1710-1721.* Virginia Historical Society Collection, New Series, Vol. I, II. Richmond, 1882-1885.

St. John, J. Hector (de Crevecoeur), *Letters from an American Farmer.* London, 1782.

Sutcliff, Robert, *Travels in Some Parts of North America, 1804, 1805, 1806.* Philadelphia, 1812.

Swedenborg. Emanuel, *Regnum Subterraneum sive Minerale de Ferro.* Dresden, 1734.

Thomas, Gabriel, *Historical and Geographical Accounts of the Province and County of Pennsylvania and West Jersey in America.* London, 1694. Reprinted New York, 1848.

Thwaites, Reuben Gold (Ed), *Early Western Travels, 1748-1846.* Cleveland, 1904-1907.

Wansey, Henry, *An Excursion to the United States of North America, 1794.* Salisbury, 1798.

Washington, George, *Writings of.* Edited by Jared Sparks. 12 vols. Boston, 1837.

Weld, Isaac, *Travels through the States of North America and the Provinces of Upper and Lower Canada during the Years 1795, 1796 and 1797.* London, 1799.

Whitworth, Sir Charles, *State of the Trade of Great Britain in its Imports and Exports Considered from 1697.* London, 1776.

Winterbotham, W. *An Historical, Geographical, Commercial and Philosophical View of the American United States.* 4 vols. London, 1795.

Winthrop, John, *History of New England.* 2 vols. Boston, 1853.

Yarranton, Andrew, *England's Improvement by Sea and Land.* London, 1677.

Zeisberger, David, *History of the Northern American Indians.* Columbus, 1910.

Newspapers and Periodicals

(a) English

American Daily Advertiser, 1791-1800.

American Museum. 12 vols. 1787-1792.

American Weekly Mercury, 1719-1742.

Carey Clippings. A collection of newspaper clippings in several volumes. Unfortunately, most are without dates and names of periodicals and are only partly classified under subject headings. Ridgeway Branch of Philadelphia Library.

Columbian Magazine, 1787.

Cramer's Pittsburgh Almanack, 1812.

Fayette Gazette, 1794-1800.

Freeman's Journal, 1783.

Hazard's Register of Pennsylvania. Edited by Samuel Hazard, vols. 1-16.

Independent Chronicle and Universal Advertiser. 1776-1784, 1784-1786, 1795-1797.

Independent Gazetteer, 1782-1797.

Independent Ledger, 1778-1786.

Lancaster Intelligencer, 1799-1800.

New Jersey Gazette, 1777-1786.

New York Gazette and Weekly Mercury, 1773-1775, 1777-1783.

New York Gazetteer (Rivington's), 1775-1778.

New York Journal or General Advertiser, 1767.

Pennsylvania Chronicle, 1767-1774.

Pennsylvania Evening Herald, 1785-1788.

Pennsylvania Evening Post, 1775-1784.

Pennsylvania Gazette, 1728-1796.

Pennsylvania German (Penn Germania), vols. I-XII. Lebanon, 1900-1914.

Pennsylvania Journal, 1773-1775.

Pennsylvania Journal or the Weekly Advertiser, 1742-1797.

Pennsylvania Ledger, 1775-1778.

Pennsylvania Magazine or American Monthly Museum, 1775-1776.

Pennsylvania Mercury and Universal Advertiser, 1784-1791.

Pennsylvania Packet and Daily Advertiser, 1771-1797.
Philadelphia Federal Gazette, 1793-1796, 1798-1800.
Philadelphia Weekly Mercury, 1720-1791.
Philadelphia Repertory, 1810.
Pittsburgh Gazette, 1786-1805.
Prime, Alfred C., *Prime Collection of Colonial Advertisements*. Fairmount Park Museum Library.
Reading Weekly Advertiser, 1796-1800.
Tree of Liberty, 1800-1803.
Union, The, 1800.
United States Gazette, 1789-1800.
Universal Asylum and Columbian Magazine, 1791.

(b) German

Chestnuthiller Wochenschrift, 1790-1794.
Deutsche Porcupein, 1798.
Geisliches Magazien, 1764-1774.
General Post Bothe, 1790.
Germantauner Zeitung, 1785-1793.
Harrisburg Zeitung, 1799-1800.
Hoch-Deutsch Pennsylvanische Geschicht-Schreiber, 1739-1752.
Lancaster Correspondent, 1799-1800.
Lancastersche Zeitung, 1752-1753.
Pennsylvanische Correspondenz, 1797-1800.
Pennsylvanische Gazette, 1779.
Pennsylvanische Staatsbote (Miller's), 1762-1779.
Pennsylvanische Staats Courier, 1777-1778.
Pennsylvania Zeitung, 1778.
Pennsylvanische Zeitungsblat, 1778.
Philadelphische Correspondenz, 1781-1800.
Philadelphische Gazette, 1779.
Philadelphisches Magazin, 1798.
Philadelphische Staatsregister, 1779-1781.
Philadelphische Zeitung, 1732, 1755-1757.
Reading Adler, 1796-1800.
Readinger Zeitung, 1789-1805.
Westliche Correspondenz, 1797-1800.
Wöchentliche Philadelphische Staatsbote, 1763-1764, 1768-1769, 1773-1774.

Publications of Historical and Other Societies

American Geographical Society, *Journal*. Vol. XXIII. New York, 1891.
American Historical Association, Annual Reports, 1889-1931. Washington, 1890-1936. The volume for 1892 contains copies of a number of tracts, found in the Bodleian Library, Oxford, relating to the struggle over the control of the colonial iron industry.
American Historical Association, *American Historical Review*. Vol. XVIII, New York, 1912.

American Institute of Mining Engineers, *Bulletin* No. XXIV. New York, 1908.

American Museum of Natural History, *Bulletin* No. VIII. New York.

American Philosophical Society, *Transactions.* Vols. I-VI, Philadelphia, 1769-1804.

American Society of Mechanical Engineers, *Transactions.* Vol. II (1881). Second Edition. New York, 1892.

Bucks County Historical Society *Papers.* 6 vols. Doylestown, 1908-1932.

Connecticut Historical Society, *Collections.* 24 vols. Hartford, 1860-1932.

Essex Institute, *Historical Collections.* Vols. I-LXVII. Salem, 1859-1932.

Geological Society of America, *Bulletin* No. II. Rochester, 1900.

Lancaster County (Pennsylvania) Historical Society, *Historical Papers and Addresses.* 37 Vols. Lancaster, 1897-1933.

Lebanon County (Pennsylvania) Historical Society, *Addresses and Papers.* 8 vols. Lebanon and Annville (Pa.), 1898-1924.

Maine Historical Society, *Collections.* First Series, Vols. I-X; Second Series, Vols. I-XXIV; Third Series, Vols. I-II. Portland, 1847-1916.

Maryland Historical Society, *Maryland Historical Magazine.* Vols. I-XXVII. Baltimore, 1906-1932.

Massachusetts Colonial Society, *Publications (Transactions and Collections).* 26 vols. Boston, 1892-1927.

Massachusetts Historical Society, *Collections.* 7 series, 77 vols. Boston, Cambridge, 1792-1927.

Massachusetts Historical Society, *Proceedings,* 1791-1930. 63 vols. Boston, 1829-1931.

New Hampshire Historical Society, *Collections.* Vols. I-XII. Concord, etc., 1824-1928.

New Haven Historical Society, *Papers.* Vols. I-VII. New Haven, 1865-1908.

New Jersey Historical Society, *Collections.* 7 vols. Newark, 1848-1872.

New Jersey Historical Society, *Proceedings.* First series, 10 vols. Newark, 1847-1867; Second Series, 13 vols. Newark, 1867-1895; New Series, 16 vols. Newark, 1916-1936.

New York Historical Society, *Collections.* 37 vols. New York, 1868-1905.

Pennsylvania German Society, *Proceedings and Addresses.* Vols. I-XLV. Lancaster, 1891-1935.

Pennsylvania, Historical Society of. *Bulletin.* Philadelphia, 1848.

Pennsylvania, Historical Society of. *Collections.* Philadelphia, 1853.

Pennsylvania, Historical Society of. *Memoirs.* 14 vols. Philadelphia, 1826-1895.

Pennsylvania, Historical Society of. *Pennsylvania Magazine of History and Biography.* 61 vols. Philadelphia, 1877-1937.

Pennsylvania Society of the Colonial Dames of America, Publications III. *Forges and Furnaces in the Province of Pennsylvania.* Philadelphia, 1914.

Rhode Island Historical Society, *Collections*. Vols. I-XXIV. Providence, 1827-1931.
Rhode Island Historical Society, *Proceedings*. New Series. Vols. I-VIII. Providence, 1893-1908.
Royal Society of London, *Philosophical Transactions*, Vol. XI. London, 1809.
Smithsonian Institute, *Annual Report*, 1903. Washington, 1903.
South Carolina Historical Society, *Collections*. 5 vols. Charleston, 1857-1897.
Virginia Historical Society, *Collections*. 12 vols. Richmond, 1833-1892.
Virginia Historical Society, *Virginia Magazine of History and Biography*, 39 vols. Richmond, 1894-1930.

General Secondary Works

Alden's Appeal Record, Robert Coleman and George Dawson *vs.* Clement B. Brooke and Henry P. Robinson, Robert W. Coleman and William Coleman. July Term, 1856. Equity. Reported 93, Pennsylvania 182. Philadelphia, 1878.
Alexander, J. H. *Report on the Manufacture of Iron Addressed to the Governor of Maryland*. Annapolis, 1840.
Ashton, T. S. *Iron and Steel in the Industrial Revolution*. London, 1924.
Bancroft, George, *History of the United States*. 6 vols. New York and London, 1916.
Bancroft, H. H. *History of the Pacific States: Mexico*. Vol. VI. San Francisco, 1883.
Bathe, G. and D. *Oliver Evans*. Philadelphia, 1935.
Bauermann, H. A. *A Treatise on the Metallurgy of Iron*. New York, 1868.
Baylies, F. *Historical Memoir of the Colony of New Plymouth*. 2 vols. Boston, 1866.
Beer, G. L. *British Commercial Policy, 1754-1765*. New York, 1907.
Beer, G. L. *The Commercial Policy of England towards the American Colonies*. Columbia University Studies III, No. 2. New York, 1893.
Beer, G. L. *The Old Colonial System*. 2 vols. New York, 1912.
Bemis, S. F. *Jay's Treaty*. New York, 1923.
Bezanson, A. Gray, R. D. and Hussey, M. *Prices in Colonial Pennsylvania*. Philadelphia, 1935.
Bining, Arthur C. *British Regulation of the Colonial Iron Industry*. Philadelphia, 1933.
Bishop, J. L. *A History of American Manufactures from 1608-1860*. 3 vols. Philadelphia and London, 1866, 1868.
Bolles, A. S. *Pennsylvania, Province and State*. Philadelphia, 1899.
Bowen, Eli, *Pictorial Sketch Book of Pennsylvania*. Philadelphia, 1854.
Boyer, C. S. *Forges and Furnaces of New Jersey*. Philadelphia, 1931.
Bruce, Kathleen, *Virginia Iron Manufacture in the Slave Era*. New York, 1931.
Bruce, P. A. *Economic History of Virginia in the Seventeenth Century*. 2 vols. New York, 1895.

Burchard, E. F. and Davis, H. W. *Iron Ore, Pig Iron and Steel.* Mineral Resources of the United States, 1924. Department of Commerce, Washington, 1928.

Burke, J. B. *Genealogical and Heraldic History of the Peerage and Baronetcy.* London, 1913.

Carter, W. C. and Glossbrenner, A. J. *History of York County, 1729-1834.* York, 1834. New edition, Harrisburg, 1930.

Chapin, H. M. *Early American Signboards.* Providence, 1926.

Chase, C. *Metallurgy of Iron.* Fort Monroe, Virginia, 1883.

Channing, Edward. *History of the United States.* 6 Vols. New York, 1926.

Clark, V. S. *History of Manufactures in the United States, 1607-1860.* Carnegie Institute of Washington, D. C., 1916. New edition, 1607-1928. 3 Vols. New York, 1929.

Dahlinger, C. W. *Pittsburgh—A Sketch of Early Social Life.* New York, 1916.

Day, Sherman. *Historical Collections of the State of Pennsylvania.* Philadelphia, 1843.

Deane, W. R. *Genealogical Record of the Leonard Family.* Boston, 1851.

Earle, Mrs. A. M. *Stage Coach and Tavern Days.* New York, 1901.

Eckel, Edwin C. *Iron Ores, their Occurrence, Valuation and Control.* New York, 1914.

Ege, T. P. *History and Genealogy of the Ege Family.* Harrisburg, 1911

Egle, William H. *History of Pennsylvania.* Philadelphia, 1880.

Evans, P. D. *The Holland Land Company.* Buffalo, 1924.

Fackenthal, B. F., Jr. *The Durham Iron Works.* Holicong, Penna., 1932.

Felt, J. B. *Annals of Salem.* 2 vols. Salem, 1849.

Firth, C. H. and Chance, J. F. *Diplomatic Relations of England with the North of Europe.* Oxford, 1913.

Flower, P. W. *A Short History of the Trade in Tin.* London, 1880.

Fonseca, F. de. and Urrutia, D. C. *Historia General de Real Hacienda,* 1845-1853. 6 Vols.

French, B. F. *The History of the Rise and Progress of the Iron Trade in the United States, 1621-1857.* New York, 1858.

Geiser, K. F. *Redemptioners and Indentured Servants in the Colony and Commonwealth of Pennsylvania.* 1901.

Goodyear, B. K. *Blast Furnaces of Cumberland County,* Pamphlet. Paper read before Hamilton Library Association, Carlisle, Penna. 1903.

Gordon, Samuel G. *Mineralogy of Pennsylvania.* Philadelphia, 1922.

Hanna, Mary Alice. *Trade of the Delaware District before the Revolution.* Smith College Studies, 1917.

Hazard, S. *Annals of Pennsylvania from the Discovery of the Delaware, 1609-1682.* Philadelphia, 1850

Hazard, S. *United States Commercial and Statistical Register,* 6 vols. Philadelphia, 1839-1842.

Herrick, C. A. *White Servitude in Pennsylvania.* Philadelphia, 1926.

Holcomb, W. P. *Pennsylvania Boroughs.* Johns Hopkins University Studies in Historical and Political Science, Vol. IV. Baltimore, 1886.

Holmes, A. *American Annals.* 2 vols. Cambridge, 1805.

Hopkins, J. C. *Cambria-Silurian Limonite Ores of Pennsylvania.* Geological Society of America, Bulletin, II, 1900.

Hunter, F. W. *Stiegel Glass.* Boston, 1914.

James, Mrs. Thomas Potts, *Memorial of Thomas Potts, Jr.* Cambridge, 1874.

Jenkins, F. *Pennsylvania, Colonial and Federal.* 3 vols. Philadelphia, 1903.

Johnson, Amandus. *The Swedish Settlements on the Delaware, 1638-1664.* 2 vols. Philadelphia, 1911.

Johnson, Emory and Others. *History of the Domestic and Foreign Commerce of the United States.* 2 vols. Washington, 1915.

Karsten, C. J. B. *Handbuch der Eisenhuttenkunde.* 5 vols. Berlin, 1841.

Keith, Charles P. *Chronicles of Pennsylvania.* Philadelphia, 1917.

Kovar, R. *Social Life in Fayette County.* Typewritten Master's thesis University of Pittsburgh.

Lesley, J. P. *Brown Hematite Iron Ore Banks in Huntingdon and Centre Counties, Pennsylvania.* Philadelphia, 1874.

Lesley, J. P. *Iron Ores of the Cumberland Valley.* N. P. 1873.

Lewis, Alonzo. *History of Lynn.* Boston, 1844.

Lewis, Alonzo, and Newhall, J. R. *History of Lynn.* Boston, 1865.

Linn, J. B. *History of Centre and Clinton Counties, Pennsylvania.* Philadelphia, 1883.

Lord, Eleanor L. *Industrial Experiments in the British Colonies of North America.* Johns Hopkins University Studies in Historical and Political Science. Extra volume XVII. Baltimore, 1898.

Macfarlane, John J. *Manufacturing in Philadelphia, 1683-1912.* Philadelphia, 1912.

McMaster, John B. *History of the People of the United States.* 8 vols. New York, 1914.

Mease, James P. *Archives of Useful Knowledge.* 3 vols. Philadelphia, 1811-1813.

Mercer, H. C. *The Bible in Iron.* Doylestown, 1914.

Mercer, H. C. *The Decorated Stove Plates of the Pennsylvania Germans.* Doylestown, 1899.

Moorehead, W. K. *Hematite Implements of the United States.* Andover, Massachusetts, 1912.

Moorehead, W. K. *Prehistoric Implements.* Cincinnati, Ohio, 1898.

Mushet, David. *Papers on Iron and Steel.* London, 1840.

Nevin, R. P. *Les Trois Rois.* Pittsburgh, 1888.

O'Callaghan, E. B. *History of New Netherland.* 2 vols. New York, 1843.

Omwake, John. *The Connestoga Six-Horse Bell Teams of Eastern Pennsylvania.* Cincinnati, 1930.

Osborn, H. S. *The Metallurgy of Iron and Steel.* Philadelphia, 1869.

Osgood, H. L. *The American Colonies in the Eighteenth Century.* 4 vols. New York, 1924.

Overman, Frederick. *Manufacture of Iron in all its Branches*. Philadelphia, 1850.

Oviedo y Valdes, Gonzalo Fernandez. *Historia General y Natural de Las Indias*. 4 vols. Madrid, 1851.

Palfrey, J. G. *History of New England*. 5 vols. Boston, 1859-1890.

Pattee, W. S. *History of Old Braintree and Quincy*. Quincy, 1878.

Pearce, Stewart. *Annals of Luzerne County*. Philadelphia, 1860.

Pearse, J. B. *Concise History of the Iron Trade of the American Colonies up to the Revolution and of Pennsylvania until the Present Time*. Philadelphia, 1876.

Pennsylvania Geological Survey—First. Made by Henry D. Rogers, State Geologist. Annual Reports, 1836-41. Harrisburg, 1836-41.

Pennsylvania Geological Survey—Second. 1874-1884. Arranged by counties. Harrisburg, 1874-1884.

Pennsylvania Society of the Colonial Dames of America. Publications III. *Forges and Furnaces in the Province of Pennsylvania*. Philadelphia, 1914.

Pennypacker, S. W. *Henry Pennebecker*. Philadelphia, 1894.

Percy, John. *Metallurgy*. London, 1864.

Peters, Richard. *Two Centuries of Iron Smelting in Pennsylvania*. Chester, 1921.

Pitkin, Timothy. *Statistical View of the United States*. New York, 1817.

Prescott, W. H. *History of the Conquest of Mexico*. 3 vols. Philadelphia, 1873.

Prescott, W. H. *History of the Conquest of Peru*. 2 vols. Philadelphia, 1902.

Priestly, H. I. *The Coming of the White Man, 1492-1848*. New York, 1929.

Ringwalt, J. L. *Development of Transportation Systems in the United States*.

Root, W. T. *Relations of Pennsylvania with the British Government*, Philadelphia, 1914.

Rupp, I. D. *History and Topography of Dauphin, Cumberland, Franklin, Bedford, Adams, Perry, Somerset, Cambria and Indiana Counties*. Lancaster, 1848.

Rutledge, J. J. *Clinton Iron Ores of Stone Valley, Huntingdon County, Pennsylvania*. American Institute of Mining Engineers, *Bulletin 24, Transactions*. 1908. New York, 1908.

Scharf, J. T. and Westcott, Thompson. *History of Philadelphia, 1609-1884*. 3 vols. Philadelphia, 1884.

Scrivenor, Harry. *Comprehensive History of the Iron Trade*. London, 1841, 1856.

Smiles, Samuel. *Industrial Biography*. Boston, 1864.

Sonn, A. H. *Early American Wrought Iron*. 3 vols. New York, 1928.

Spencer, A. C. *Magnetite Deposits in Berks and Lebanon Counties*. Bulletin 315. United States Geological Survey. pp. 185-189, 1907.

Spencer, A. C. *Magnetite Deposits of the Cornwall Type in Pennsylvania*. Bulletin 359, United States Geological Survey, 1908.

Swank, J. M. *Introduction to a History of Ironmaking and Coal Mining in Pennsylvania.* Philadelphia, 1878.

Swank, J. M. *History of the Manufacture of Iron in all Ages and Particularly in the United States from Colonial Times to 1891.* 2nd Edition. Philadelphia, 1892.

Swank, J. M. *Progressive Pennsylvania.* Philadelphia, 1908.

Thacher, J. B. *Christopher Columbus, His Life, His Work, His Remains.* 3 vols. New York, 1903-1904.

Trumbull, Benjamin. *Complete History of Connecticut.* 2 vols. New Haven, 1818. New edition, New London, 1898.

Turner, E. R. *Negro Slavery in Pennsylvania.* Washington, 1910.

United States *Department of Commerce Publications, 1924.* Washington, 1925.

United States, *Geological Survey Bulletins.* Nos. 315, 359, 491. Washington, 1907, 1908, 1911.

Ure, Andrew. *Dictionary of Arts, Manufactures and Mines,* 2 vols. New York, 1857.

Wallace, P. B. *Colonial Ironwork in Old Philadelphia.* New York, 1930.

Watson, John F. *Annals of Philadelphia (and Pennsylvania).* 3 Vols. Philadelphia, 1881.

Weeden, W. B. *Economic and Social History of New England, 1620-1789.* 2 vols. Boston, 1890.

Westcott, Thompson. *Life of John Fitch.* Philadelphia, 1857.

Whitehead, W. A. *East Jersey under the Proprietary Governments.* Newark, N. J., 1875. Appendix contains a reprint of George Scot, *Model of the Government of the Province of East New Jersey in America,* 1685.

INDEX

A

Abbington Iron Works, 53, 187.
Acrelius, Israel, 41, 69, 141.
Agreements, wage, 118-119.
Agricola, G., 91.
Agriculture, 35.
Air furnaces (early cupolas), 55, 87-88, 99, 181, 182.
Alden, M. F., 189.
Alleghenies, 29, 35, 62, 92.
Allen, William, 54, 57, 132, 136.
Alliance Iron Works (Jacob's Creek Furnace), size of plantation, 31; legend, 43-44; established, 64, 190; partners, 132, 171; failure of, 141.
Alsace, 131.
American Iron Company, 109.
American Philosophical Society, 98, 100.
Anchor works, 55, 56.
Anchors, 183.
Ancony, 84, 123.
Anne, Queen, 132.
Anshutz Furnace, See Huntingdon Furnace; Pittsburg Furnace.
Anshutz, George, 61, 131, 190.
Anthracite coal measures, 69.
Antietem Forges, 190.
Apprentices, 111, 113, 117-118.
Armstrong, Joseph, 59.
Arthur, John, 59.
Ashman, George, and Company, 60, 189.
Associations, ironmasters, 145-146.
Aztecs, 12.

B

Backhouse, Richard, 136.
Ball Bloomery, 187.
Ball, John, 53, 187.
Baltic countries, 150, 183.
Baltimore, 36, 40, 41.
Bantling, Emanuel, 105.
Barbadoes, 20.
Barree Forge, 61, 191.
Bar iron, forged, 34; transported to frontier, 39-40, 61; made on iron plantations, 45; manufactured into tools and implements, 53; uses on frontier, 63; English, 73; output of, 85; bloomery, 121; duties on, 149, 150, 152, 153, 154; Swedish, 154; enumerated, 157; Pennsylvania tariff regarding, 161; import, 164, 183; reports on, 178; prices of, 178.

Barr, John, 57, 141.
Barter, 173, 174.
Beale, Thomas, and Company, 61, 190.
Bedford, 61, 183.
Bedford Forge, 190.
Bedford Iron Works, 60, 189.
Bellefonte Forge, 61, 191.
Bellows, bloomery, 76; blast furnace, 78-79, 82; forge, 82, 84-85.
Benezet, Daniel, 142.
Benner, General Philip, 61, 139, 190, 192.
Bennett, William, 59, 188.
Berkeley, John, 16.
Berkshire Furnace, 180, 188.
Bessemer process, 92.
Bethlehem, 36, 56.
Bethlehem Steel Corporation, 58.
Billington Iron Works, See Turner's Iron Works.
Bills of credit, 36, 173, 174, 175.
Bird, Mark, inherits ironworks, 51; enterprises with James Wilson, 55, 134, 171; uses concave rolls, 105; member of Assembly, 135-136; in Revolutionary army, 139; bankrupt, 141, 143; slitting mill, 159; ironworks of, 189; agreement, 194-196.
Bird, William, 51, 124, 170, 188.
Birdsboro Slitting Mill and Steel Furnace, 51, 136, 143, 189.
Birmingham, 149, 150.
Bituminous coal measures, 69.
Blacksmiths, in Philadelphia, 24; work iron ores, 26; on plantations, 31; on frontier, 63; artisans in iron, 84, 85, 114; scarcity of, 108; wages, 121; number of, 182; in towns, 183.
Blast furnaces, 15; description of, 33; contrast early with modern, 46; height of, 77, 193; processes, 77-82; bellows, 78-79; production, 119; cost of erection, 170; number of, 182, 183.
Blister steel, See Steel.
Blockley and Merion Society for Promoting Agriculture and Rural Economy, 35.
Bloomeries, early, 18, 26, 63; bellows, 63; description of processes, 76-77; workers in, 121; attempts to suppress in colonies, 153; number of, 183.
Bloomery forge, See Bloomeries.
Blowing tubs, 82, 83, 95-96.

Board of Trade, 150-151, 155, 157, 174.
Boiling Springs Iron Works, See Carlisle Iron Works.
Boker, George H., 42.
Boon (Boone), G., 170.
Boroughs, 38-39.
Bosh, 78, 79.
Boston, 174.
Boulton, Matthew, 99, 102.
Bounties, 150, 158.
Boyer, John, 118.
Braintree, 17.
Brandywine, Battle of the, 42, 139.
Brandywine Slitting Mill, 45.
Branson's Steel Furnace, 55, 188.
Branson, William, partner in Coventry Iron Works, 51; steelmaker, 55; at Windsor Forge, 56; seeks labor in Europe, 108; nationality, 131; merchant, 132; Ironworks of, 187, 188.
Bridges, iron, 103.
Bridges, Robert, 16.
Briggs, Samuel, 104.
Bringewood, 82.
Bristol, 38, 149, 150.
Broadcloth, 34.
Brobst, John, 192.
Brooke Forge, See Buckley's Forge.
Brown hematite ores, 68, 69.
Brown, William, 61, 191.
Buckley, Daniel, 189.
Buckley's Forge, 189.
Bull, John, 139.
Bull, Thomas, 137.
Burkhart's Forges, See Antietem Forges.
Byberry Plating Mill, 188.
Byrd, William, 150, 152.

C

Caldwell, David, 57, 188.
Caledonia Forge, 61, 192.
Canada, 109.
Cannon, in seventeenth century New England, 19; cast at Warwick and Reading, 42; cast at Mary Ann Furnace and Codorus Iron Works, 59; attempts to forge, 60; cast in southeastern Pennsylvania, 112, 180; imported from England, 179; cast at air furnaces, 181; exports of, 183.
Capital, in early Massachusetts, 17; lack of, 169, 172; English, 171; Dutch, 171; merchant, 172.
Capitalistic enterprises, 107-108, 109.
Carbonate ores, 68, 69.

Cards, cotton and wool, 104, 126, 183.
Carlisle, 39, 59, 60, 139.
Carlisle Armory, 60, 180, 181.
Carlisle Iron Works, 59, 170, 188.
Carlisle Rolling and Slitting Mill and Steel Furnace, 189.
Carpenters, 25.
Carriage wheel boxes, 99.
Carron Iron Works, 73, 83.
Carters, See Teamsters.
Castings, process of, 33; variety of, 34, 80; air furnace, 99; wages for work on, 119-120; duties on, 164; prices, 178; exports of, 183.
Cast iron, description of, 91.
Catalan forge, 76, 77.
Census of 1810, 183.
Centre Furnace, 61, 190, 193.
Chafery, 84.
Chambers, Benjamin, 60, 139.
Chambers, George, 60.
Chambers, James, 60, 139, 190.
Chambers, William, 60.
Chambers, W., and Brothers, 189.
Chapman, John, 121.
Charcoal, 31, 70-75.
Charcoal burners, See Colliers.
Charles II, 22.
Charlotteburg, 110.
Charming Forge, production of bar iron, 85; Hessian prisoners of war employed, 112; Negroes employed, 115; purchased by Henry William Stiegel, 141; George Ege at, 142; slitting mill built, 159; origin, 188.
Charming Forge Slitting Mill, 113, 159, 189.
Cheltenham Slitting Mill, 52, 192.
Child labor, 111, 115, 126-127.
Christiana River, 41.
Christina, 41, 69.
Christina, Queen, 22.
Churches, 37.
Cinder iron, 95-96. See also Slag.
Civil War, 29, 58.
Clifford, Thomas, Jr., 104, 132.
Clinton ores, 68.
Coalbrookdale Iron Works (England), 72. See also Colebrookdale.
Codorus Iron Works, 37, 59, 136, 141, 180, 188.
Coke, 72-73, 92.
Coins, 174.
Colebrook Furnace, legend, 42-43; origin, 58, 133, 190; height and size of, 193; iron made at, 194.

Colebrookdale Furnace, cost of transporting iron from, 39; named after Abraham Darby's Shropshire furnace, 50; origin, 51; abandoned, 53; Indian labor at, 117; damages to, 120-121; ownership, 144, 170, 187.

Coleman, Robert, ironworks, 58, 170, 190; nationality, 131; clerk, 133; in Revolutionary army, 139; and James Old, 144.

Colliers, number of, 74; work of, 75; scarcity of, 108; agreements, 118; wages, 123-124.

Colonial agents, 150, 152.

Columbian, 101.

Columbus, Christopher, 11, 12.

Colles, Christopher, 100.

Companies, See Partnerships.

Concord, 17.

Conestoga wagons, 39.

Connecticut, 18-19.

Connellsville, 92.

Constitution, 159.

Constitutional Convention, 1787, 136.

Continental army, 42, 112.

Continental Congress, 112, 134, 136, 174, 180.

Contract labor, 110, 118.

Copper, 12, 25.

Cornwall Bloomery, 57-58.

Cornwall Furnace, size of plantation, 31; origin, 56, 57-58, 187; casting of cylinders, 96; Robert Coleman and, 133; cost of production at, 177; casting of cannon and salt pans during Revolution, 180; height and size of furnace, 193; iron made at, 193-194.

Cornwall mines, 57, 58, 68.

Cort, Henry, 88, 105.

Cortez, Fernando, 12.

Cortlandt, 110.

Coryell's Ferry, 54.

Counterfeit bills, 173-174.

Coventry Iron Works, size of plantation, 31; origin, 51, 187; scarcity of workers, 111; Samuel Savage, Jr., and, 133; owners, 144.

Coventry Steel Furnace, 45, 51, 156, 187.

Coxe, Tench, 127, 170, 181, 182.

Credit, 172-173.

Cromwell Iron Works, See Bedford Iron Works.

Cromwell, Thomas, 60-61, 189.

Crosby, John, 141.

Crowley, Sir Ambrose, 24.

Crucible steel, 86.

Crum Creek Forge, 41, 54, 141, 187.

Cumberland Furnace, 60, 191.

Cumberland Valley, 59.

Cupolas, See Air Furnaces.

Cuthbertson, John, 37.

D

Dale Furnace, 52, 190.

Damascus steel, 91.

Darby, 56.

Darby, Abraham, 50, 72.

Darby Tilt Hammer and Plating Mill, 188.

Darts, 11.

Davies, David, 121.

Declaration of Independence, 136.

Delaplank, Frederick, 141, 189.

Delaware, 22-23.

Delaware River, 40, 41, 68, 101, 182.

Delaware Valley, 53-56.

Delaware Works, 55, 134, 171, 189.

Denmark, 154.

Denny, Governor, 178.

Dewees, William, 131, 139, 194-196.

Dickinson, John, 134.

Dicks' Bloomery, 188.

Dicks, Peter, 54, 58, 188.

District Forge (No. 1), 190.

District Forge (No. 2), 191.

District Furnace, 52, 189.

Doe Run Forge, 190.

Donaldson, Arthur, 101.

Dorsey Forge, See Baree Forge.

Dorsey, G. P., 61, 191.

Dowlin, 190.

Dowlin Forge, See Mary Ann Forge.

Drinker, Elizabeth, 41.

Drinker, Henry, 71, 74, 75, 121, 170, 172.

Drunkenness, 38, 39.

Dubbs, Martin, 171.

Dudley, Dud, 72.

Dunlop, James, 61.

Dunlop, John, 61, 191.

Duquesne Forge, 57, 192.

Durham, 24.

Durham Iron Works, transportation of iron from, 40; travelers visit, 41-42; origin, 53-54; mines, 70; miners, 71; Hessian prisoners of war employed, 112; Negro workers, 115; Joseph Galloway and, 134, 136; iron exported to England from, 152; sold by Commissioner of Forfeited Estates, 170; partners, 171; casting of cannon, 180.

Dutch, 21-23.

Duties, English, at end of seventeenth century, 20; struggle among English groups regarding colonial, 149; Naval Stores Bill, 151; reduced on colonial pig iron, 152; abolished on colonial pig iron, 155; abolished on colonial bar iron, 157; after Revolution, 159-160; Pennsylvania, 160-161; under Jay Treaty, 162-163; United States, 164-165.

E

East India Company, 14.
Eckert, Valentine, experiment in manufacturing wire, 104; nationality, 131; member of Assembly, 135; member of Pennsylvania Constitutional Convention of 1776, 136; commissioner for purchasing provisions for army during Revolution, 139; ironworks of, 189, 190.
Eckert and Voigt's Wire Mill, 189.
Ege, George, hires Hessian prisoners of war, 113; member of Assembly, 136; at Charming Forge, 142, 159; ironworks of, 189, 190, 191.
Ege, Michael, 59, 60, 170, 191.
Ege, Rebecca, 144.
Eichbaum, 104.
Elizabeth Furnace, origin, 57, 188; bought by Henry William Stiegel, 57, 141; charcoal house, 72; Hessian prisoners of war employed, 112; Robert Coleman and, 133; John Dickinson and, 134; iron made at, 194.
Ellis, Robert, 132.
England, scarcity of wood in, 14; ores from, 19; bar iron from, 24; exports, 34, 179; blast furnaces in, 77; inventions in, 82-83; labor in, 107-108; demand for iron in, 150; first exports of iron from Pennsylvania to, 152; admittance of colonial iron duty free into, 153; colonial policy, 159; laws prohibiting exportation of machinery and migration of artisans, 161; iron imports from Pennsylvania, 179.
England, John, 53.
English money, 36, 173.
Enlistments, 111-112.
Etter, Gerrard, 131, 188.
Europe, silver and gold sent to, 14; feudal manors of, 31; capital from, 171.

Evans, Oliver, 101, 102, 103, 104.
Evans, Thomas, 61.
Exportation of machinery, 161.

F

Fairchance Iron Works, See Fairfield Iron Works.
Fairfield Iron Works, 64, 175, 191, 193.
Fairs, 38.
Fairview Furnace, See Mary Ann Furnace, Fayette County.
Fallen Timbers, 62.
Falling Creek Iron Works, Virginia, 16.
Feast of Roses, 142-143.
Federal Slitting Mill, See Rokeby Slitting Mill.
Fillers, 81, 121.
Fireplaces, 97-99, 100.
Flax, 35.
Finley, James, 103.
Fitch, John, 101, 104.
Fitzwater, George, 132.
Flower, Samuel, 57.
Forest Mountains, 56.
Forge, See Refinery Forge.
Forgemen, 118, 121-123.
Fort Casimir, 22.
Fort Christina, 22.
Fort Loudon Bloomery, 60.
Fort McIntosh, 63.
Fort Nassau, 21.
Foundries, See Air Furnaces.
Foulke, Caleb, 55, 191.
Foulke's Steel Furnace, 55, 191.
Founders, number required at furnaces, 81; scarcity of, 108; skilled, 113; agreements, 118; wages, 119-120.
France, 154.
Franklin, Benjamin, friend of Robert Grace, 43; inventor of a fireplace, 97-98, 99; present of model of bridge from Thomas Paine, 103; advice on investments in ironworks, 134; and bills of credit, 173.
Franklin stoves, 97-98, 99.
French Creek Forge, 190.
French, Nathaniel, 170.
French and Indian War, 60, 111-112, 137.
French loans, 141.
French money, 36, 173.
French Revolution, 103.
Freedom Forge, 61, 191.
Free Society of Traders, 24.
French money, 36.
Frontier, 40, 62, 63.
Furs, 40.

G

Galloway, Joseph, 54, 134, 136, 137.
Garrison, Nicholas, 104.
Gee, Joshua, 150.
George I, 150.
German money, 36, 173.
Germantown, 182.
Germantown Steel Furnace, 191.
Germany, 161.
Gettysburg, 59.
Gibraltar Forge, 189.
G:bson, J., and Company, 191.
Gibson, Nathaniel, 32.
Glasgow Forge, 144, 187.
Gloninger, John, 61, 171.
Glorious Revolution, 132.
Gloucester, 21.
Gold, used by Aztecs, 12; used by Incas, 13; sent to Europe, 14; sought in Pennsylvania, 25; co:ns in Pennsylvania, 36, 174, 175.
Gordon, Governor, 115.
Grace, Mrs. Robert, 37.
Grace, Robert, friend of Benjamin Franklin, 43; and Franklin fireplaces, 97-98; nationality, 131-132; at Warwick Furnace, 134; training in metallurgy, 145.
Graves, Thomas, 16.
Great Britain, See England.
Great Valley (Valley Forge) Works, 52, 189.
Greene Furnace, 64, 192.
Green Lane Forge, 115, 187.
Greenwood Furnace, 191.
Grenville, Sir Richard, 14.
Griffiths, Robert R., 171.
Gristmills, 34-35, 170.
Grubb, Colonel Peter, 139.
Grubb, Curtis, 58, 135, 139.
Grubb, Henry B., 170.
Grubb, Peter, 56, 57, 58, 187.
Grubb, Peter, Bloomery, 57-58, 187.
Grubb, Peter, Jr., 57, 58, 170.
Grubb, Peter, Jr., and Company, 189.
Gu:lds, 108, 127.
Gulf Forge, 189.
Gunsmiths, 182, 183.
Guttermen, 81, 113, 121.

H

Hall Forge and Tilt Mill, 191.
Hall, John (1), 188.
Hall, John (2), 191.
Hamilton, Alexander, 164.
Hamilton, James, 155.
Hammermen, 113, 123.
Hammers, 84, 116, 122.

Hampton Forge, 191.
Hanbury, Major, 88.
Hanover, 131.
Hariot, Thomas, 14.
Harmony Forge, 61, 191.
Harrisburg, 36, 39.
Hasenclever, Peter, 96, 109-110.
Hasenclever, Seton and Crofts, 109, 110.
Hatherly, Timothy, 17.
Hay Creek Forges, See New Pine Forges.
Hayden's Bloomery, 64.
Hayden, John, 175, 190, 191.
Hearth, 78, 79.
Helmstead Forge, 188.
Hemp, 35.
Henry VIII, 77.
Hereford Furnace, 53, 188.
Hermelin, S. G., on abandoned ironworks, 52; on Delaware Works, 55; on manufactures, 56; on mining, 70; on water wheels, 82; on plantation stores, 124-125; on cost of production, 177.
Hessians, 112.
Hibern:a Forge, 191.
Hillegas, Michael, 134.
Hiltzheimer, Jacob, 41.
Hockley, Richard, 57, 154.
Hockley, Thomas, 139.
Holker. Robert, 132, 171.
Holland Land Company, 171.
Hoops, 179, 183.
Hope Furnace, 61, 191.
Hopewell Forge (Bedford County), 192.
Hopewell Forge (Berks County), 51, 188.
Hopewell Forges (Lebanon County). 31, 133, 170, 178, 187.
Hopewell Furnace, origin, 51, 143, 189; fireplace, 100; James Wilson and, 134; height and size of, 193.
Hopkins, John, 132.
Hornblower, 100.
Horse racing, 38.
Huber Furnace, See Elizabeth Furnace.
Huber, John Jacob, 57, 141, 188.
Hudde, Andrie. 22.
Humphreys' Steel Furnace, 55, 188.
Humphreys, Whitehead, 103, 158, 180, 188.
Humphries, David, 127.
Humphries, Soloman, 188.
Huntingdon, 183.

Huntingdon Furnace, 61, 191, 193.
Huntsman, Benjamin, 86.
Huntsman process, 92.

I

Incas, 12-13.
Indentured servants, 109, 111, 114.
Indentures, 117-118.
Indians, use of oxides by, 11; contacts with Europeans, 13; Narragansett, 19; trails of, 39; in ironworks, 111, 117.
Industrial Revolution, 108.
Inns, 39.
Inventions, 95-105, 176.
Ireland, 173.
Iron, round bolt, 55, 105.
Iron Act of 1750, 55, 137, 150, 154-155, 158.
Iron exports, 152, 153, 157, 177-179, 183.
Iron manufactures, made from bar iron, 85; made by blacksmiths, 114; British, 137, 150, 153; cast, 141; American, 164, 165; used in shipbuilding, 182; Pennsylvania, 183.
Iron ores, in West Indies, 11; in Mexico and Peru, 12, 13; in North Carolina, 14; in Virginia, 14; in Massachusetts, 16; in Connecticut, 19; English, 19; in Rhode Island, 19; in New Jersey, 20; in Maryland, 20; in Delaware, 23; worked by blacksmiths, 26; types, 67; classification, 68; distribution in Pennsylvania, 68-69.
Ironmasters, nationalities, 131-132; classes, 132; merchant, 132; and agriculture, 132-133; civic and political leaders, 134-137; military, 137-139; difficulties during Revolution, 139-140; failures among, 140 - 144; successes among, 144; matrimonial alliances, 144; associations of, 145-147; British, 149, 152, 155; and removal of duties, 154; petitions from, 165; exchanges with merchants, 175.
Iron plantations, description of, 30-31.
Iron sand, 68.
Iron sulphides, 69.
Ironworks, cost of, 169-170. See also Blast furnaces, Refinery forges, Bloomeries, Slitting mills, Steel furnaces, Plating mills, Air furnaces.
Ironworkers, See Workers.

J

Jacob's Creek, chain bridge, 103.
Jacob's Creek Furnace, See Alliance Iron Works.
Jacobs, Cyrus, 57, 134, 144, 170, 190.
James, Duke of York, 22, 23, 173.
James, S. and Company, 187.
James River, 16.
Javelins, 11.
Jay Treaty, 159, 162.
Jenkins, David, 57, 131, 133, 136.
Jenks, Joseph, 19.
Jesuits' bark, 34.
Joanna Furnace, 52, 190, 191.
Jones, B. and R., 191.
Jones, Rees, 123.
Journeymen, 111, 117.
Juniata Forge, 61, 191.
Juniata Iron Company, 60.
Juniata River, 40.
Juniata Valley, 29, 39, 60-62.

K

Kalm, Peter, 35, 41.
Keepers, 81, 113, 118.
Keith Furnace, 187.
Keith, Sir William, 26, 53, 131, 132, 169, 187.
Kensington, 56.
King Philip's War, 17.
King, Rufus, 162.
Kinsey, Doctor, 101.
Knight, James, 82, 95.
Kurtz, 187.
Kurtz' Iron Works (Bloomery), 56, 187.

L

Labor, See Workers.
Laborers, 124.
Labor institutions, 107.
Labor organizations, 127-128.
Labor system, 107.
Labor unions, See Labor organizations.
Lackawanna Bloomery, 190.
Lafayette, Marquis de, 43.
Lake Superior ores, 57, 68, 92.
Lamb's Forge, See Youghiogheny Forge.
Lancaster (Massachusetts), 17.
Lancaster (Pennsylvania), 39, 182.
Lancaster Turnpike Company, 101.
Lane, William, 192.
Langhorne, Jeremiah, 132.
Lardner, Flower and Hockley, 133.
Lardner, Lynford, 35, 57, 136.
Latrobe, Benjamin H., 100, 101, 102.

Laurel Furnace, 64, 191.
Leacock, John, 170.
Lebanon, 79.
Leephar, Crotzer and Company, 191.
"Legend of the Hounds," 42-43.
Legislation, British, See Parliament.
Legislation, Pennsylvania, restricting sale of liquor, 38; regulating enlistments, 112; regarding apprentices, 117; tariff, 159-161; prohibiting exportation of machinery and migration of artisans, 161.
Leonard, Nathaniel, 17-18.
Leonard, Thomas, 17-18.
Lesher, Jacob, 139, 191.
Lesher, John, 131, 136, 139, 190.
Lewis Forge, 190.
Lewis, James, 171.
Lewis, T., 190.
Lewis, William, 61, 190, 191.
Lewistown, 61.
Liberty Forge, 190.
Licking Creek Forge, 61, 190.
Limestone, 67, 77, 80.
Limonite ores, 68, 69.
Lincoln, Mordecai, 51.
Liquors, laws regulating sale of, 38.
Little Falls (Bloomery) Forge, 32, 192.
Little Pine Forges, 144.
Liverpool, 149.
Lobingier, Christopher, 171.
Logan, James, 24, 54, 136, 151, 152, 171.
Logan Iron Works, 61, 191.
London, Peter Hasenclever and. 96, 109; American iron in, 97, 155; manufacturers of, 149, 150; Pennsylvania iron in, 183.
London Company, See American Iron Company; Virginia Company of London.
Long Pond, 110.
Lotteries, 142.
Loudon Iron Works, 60, 190.
Lower, Christian, 133, 136, 137, 139.
Lumber, 35.
Lynn, Massachusetts, 17.

M

Machinery, 35, 161, 163.
Maclay, William, 191.
Magnetic iron sand, 23.
Magnetite ores, 68.
Malbon, 16.

Managers, 113, 119.
Manatawny, 50, 51.
Manhattan Water Company, 102.
Manheim, 141, 142.
Manufacturers, See Iron manufactures.
Marcus Hook, 182.
Maria Forge, 188.
Marke, Thomas, 171.
Markets, 38.
Marmie, Peter, 43, 131, 132, 171.
Marquedant, Charles, 171.
Martic Iron Works, 57, 115, 134, 188.
Martin, J., 192.
Mary Ann Forge, 55, 190.
Mary Ann Furnace (Berks County), 52, 191, 193.
Mary Ann Furnace (Fayette County), 64, 192.
Mary Ann Furnace (York County), 59, 112, 139, 180, 188.
Maryland, 20, 153.
Masborough, 103.
Massachusetts, 16-18, 104.
Massey, Mordecai, 61.
Massey, Mordecai, and Company, 191.
Massey, Phineas, 61, 191.
Matrimonial alliances, 144.
Mattakeeset Pond, 17.
Maybury, Thomas, 187, 188.
Maybury Furnace, See Hereford Furnace.
McCall, George, 131, 136, 187.
McCall, Samuel, 131, 132.
McCall's Forge, See Glasgow Forge.
McCall's Wharf, 141.
McDermett, William, 61, 192.
Meason, Isaac, 136, 172, 190, 191.
Medicine, types of, 34; practice of, 45.
Mercersburg, 40.
Metallurgy, 144-145.
Meteors, 11, 67.
Methodists, 37.
Mexico, 12, 13, 14.
Michaux, F. A., 41.
Middletown, 40.
Mifflin, George, 170.
Milesburg Iron Works, See Harmony Forge.
Miles, Dunlop and Company, 191.
Miles, Samuel, 61, 190.
Miners, 71, 124.
Mines, 25.
Mining operations, 70.
Mochabee, Reuben, 191.
Money, scarcity of, 34, 173; foreign coins, 36, 173.

Monongahela country, early homes, 32; eastern bar iron sent to, 39-40; iron plantations, 45; industry, 49; establishment of ironworks, 62-65.
Morgan, James, 41, 131.
Morris, Anthony, 54, 132, 172, 187.
Morris, Anthony, and Company, 187.
Morris, John, 59.
Morris, Lewis, 20.
Morris, Samuel, 59.
Moselem Forge, 188.
Mounds, Indian, 13.
Mount Holly Iron Works, 60, 189.
Mount Hope Furnace, 57, 133, 170, 189, 190.
Mount Joy Forge, See Valley Forge.
Mount Joy Furnace, See Colebrook Furnace.
Mount Pleasant Forge, 144, 187.
Mount Pleasant Furnace (Berks County), 52, 144, 187.
Mount Pleasant Iron Works (Franklin County), 60, 189.
Mount Vernon Iron Works, 64, 191.
Murray, Griffin and Bullard, 181.

N

Nail machines, 104-105.
Nailmakers, 182-183.
Nails, scarcity on frontier, 63; attempted restrictions on manufacture of, 151, 153; prices of, 178; progress of manufacture, 182; exports of, 183.
Nail works, 56, 88, 113, 117, 170, 183.
Nancarrow, John, 99.
Nancarrow and Matlack's Steel Furnace, 55, 190.
Nanticoke Bloomery, 189.
Naval stores, 151, 152, 159, 162.
Naval Stores Bill, 151, 152.
Negro workers, blacksmiths, 26; slaves and freed, 111; duties on importation of, 114-115; relations between ironmasters and, 118; teamsters, 124.
New Amsterdam, 22.
New Castle, 22, 182.
Newcastle (England), 24.
New England, early iron industry in, 16-20; exports of rum, 34; bloomeries, 77; first steam engines in, 99-100; forges use Pennsylvania pig iron, 179.
New Haven, 19.

New Jersey, first attempts to produce iron in, 20; mines in, 70; enterprises of Peter Hasenclever in, 96, 109; Schuyler Copper Mines, 100; Soho Works, 102; labor, 109; meeting of ironmasters, 145.
New London, See Pequot.
Newmarket Forge, See Quittapahilla Forge.
New Netherland, 21, 22, 23.
New Pine (Hay Creek) Forges, 51, 188.
New Sweden, 22.
New York, 36, 96, 100, 102, 109, 110.
Nicholson, John, 137, 190.
Nikoll, J. G., and Company, 188.
Nixon, John, 143.
Non-importation agreements, 137.
North Carolina, 14.
Northern War, 150.
Nutt, Anna, and Company, 111, 187.
Nutt, Samuel, Sr., begins iron manufacture, 31, 53; Emanuel Swedenborg visits, 40-41; a pioneer ironmaster, 43; early career, 51; seeks skilled labor in Europe, 108; nationality, 131; member of Assembly, 134; ironworks of, 187.

O

Octorara region, 57.
Offley, Daniel, 55, 188.
Offley's Anchor Forge, 55, 188.
Ohio River, 62, 63.
Old, Davies, 170.
Old, James, at Speedwell and Poole Forges, 57; nationality, 131; justice of the peace, 136; bankrupt, 141; family, 144; ironworks, 188, 189.
Old, Joseph, 144.
Old, William, 144.
Oldmixon, John, 182.
Oley (Spang) Forge, 188.
Oley Iron Works, established, 51; iron mines, 70; charcoal consumed, 75; Daniel Udree and, 137, 138; casting of cannon, 180; height and size of furnace, 193.
Oliphant, J., 191.
Oliphant's Forge, See Sylvan Forge, 64.
Open hearth process, 92.
Orders-in-Council, 162.
Ore burners, 121.
Ores, See Iron ores.
Orukter Amphibolos, 101.
Oxygen, 13, 68.

P

Paddington, 103.

Paine, Thomas, inventor, 103.

Paine, William, 19.

Paper currency, See Bills of credit.

Paris, Treaty of (1783), 52, 159, 176, 181.

Parliament, grants drawbacks on iron, 20; Iron Act of 1750, 55, 137, 150, 154-155, 158; restricts use of timber, 71; laws regarding colonial enlistments, 112; prohibits trade with Sweden, 150; considers Naval Stores Bill, 151; proposed restrictions on colonial iron manufactures, 153; and European diplomacy, 154; enumeration of iron, 157, 179; report regarding ironworks to, 178; prohibits exportation of machinery and migration of artisans, 161.

Partnerships, 170-171.

Paschall's Steel Furnace, 55, 188.

Paschall, Stephen, 55, 188.

Patent laws, 105.

Patton and Bird, 118.

Patton, Colonel John, 61, 139, 190.

Patton, John, 136.

Payne, John, 88.

Pears' Bloomery, See Plumsock Forge.

Pears' Slitting Mill, 64, 192.

Pears, Jeremiah, 191, 192.

Peirsol, Mordecai, 188.

Penn, Thomas, 109, 154.

Penn, William, ironmaster at Hawkhurst, 23; encourages iron manufacture in Pennsylvania, 24; first settlement in Pennsylvania, 26; and Joshua Gee, 150.

Pennsylvania Constitutional Convention (1776), 136.

Pennsylvania Constitutional Convention, 1790, 136-137.

Pequot (New London), 19.

Percy, John, 91.

Peru, 12, 13.

Peruvian bark, 34.

Pewter dishes, 33.

Philadelphia, blacksmiths, 24; imports, 36; stage coaches, 36; trade, 39, 40, 175; steel furnaces, 55, 56; tin plate works, 56; nail works, 56, 88; wire mills, 89; sale of Franklin stoves in, 97; street water pipes, 102; rolling and slitting mill, 102; steam engines, 103; wire mills, 104; prices, 124, 125; production of cotton and wool cards, 126; merchants, 132; McCall's Wharf,

141; town meeting on tariff, 160; stores, 171; currency in, 174; prices of iron in, 177; contracts for arms, 181; shipbuilding, 182; number of blacksmiths, 183; exports, 183.

Philadelphia Society for Promoting Agriculture, 35.

Phoenixville Slitting Mill and Nail Works, 190.

Pig iron, product of blast furnace, 34, 80; transportation of, 39; made at plantation ironworks, 45; production in Schuylkill Valley, 53; English, 73, 149; wages for casting, 119; British duties on, 150, 152, 153, 154, 155; suggested bounties, 151; enumerated, 157; Pennsylvania duties on, 161; policy of Great Britain after Revolution regarding, 162; American duties on, 164; prices of, 178; exports, 179, 183.

Pine Forge, 115, 187.

Pine Forge Slitting Mill, 159, 189.

Pine Grove Forge (Fayette County), 64, 190.

Pine Grove Forge (Lancaster County), 57, 192.

Pine Grove Furnace, 59, 188, 193.

Pittsburgh, early, 45; gateway to west, 61; wire mill, 90; labor organizations in, 128; merchants, 132; manufacturing in, 183.

Pittsburgh (Anshutz) Furnace, 64, 190.

Plating mills, locations of, 45; processes, 88; workers at, 113, 117; restrictions imposed by Parliament on, 155; cost of, 170; erected during Revolution, 181; number of, 183.

Plumsock Forge, 64, 191.

Plumsted, Clement, 132.

Plymouth, 171.

Pool Forge, 144, 170-171, 187.

Poole Forge, 57, 134, 170, 189.

Portuguese money, 36.

Potters, 81, 82, 119-120, 121.

Potts, David, 137.

Potts and Hobart, 190.

Potts, John, Sr., 132, 136, 188.

Potts, John, 137.

Potts, Isaac, 52, 139.

Potts, Isaac, and Company, 189.

Potts, Joseph, 52, 159, 189, 194-196.

Potts, Samuel, 135, 136, 159, 189, 190.

Potts, Samuel, and Company, 158, 188, 190.

Potts, Thomas (1), 43, 131, 170, 187.
Potts, Thomas (2), 135, 136, 137, 139, 145.
Potts, Thomas, and Company, 158, 188.
Potts, Thomas, Jr., and Company, 187.
Pottstown, 145.
Potts, William, 189.
Pottsgrove Forge, 115, 188.
Pottsgrove Steel and Air Furnaces, 158, 188.
Preston, Samuel, 170.
Prices, 125-126, 177-178.
Priestley, Joseph, 145.
Principio Company, 53.
Principio Iron Works, 20, 21.
Printz, Johan, 22, 23.
Probst, John, 41, 131.
Probst's Works, See Westmoreland Iron Works.
Providence, 19.
Provincial Conference (1776), 136, 180.
Pueblo region, 13.

Q

Quakers, 21, 112.
Quittapahilla Forge, 58, 133, 188.

R

Raleigh, Sir Walter, 14.
Randolph, Edward, 19.
Rawle, Francis, 136, 171.
Raynham Iron Works, Massachusetts, 17.
Read, Charles, 132.
Reading, 36.
Reading Forge, 190.
Reading (Redding) Furnace, John Cuthbertson at, 37; George Washington at, 42; established, 51, 187; rivals English furnaces, 51; abandoned, 52; mines, 70; length of blast, 82; casting of cannon, 187.
Reading Furnace (Berks County), 190.
Reaumur, R. A. F. de, 91.
Rebecca Furnace, 188.
Redemptioners, See Indentured Servants.
Red hematite ores, 68.
Redstone Furnace, 64, 191, 193.
Reese, L., and Company, 191.
Refinery forges, description of, 18, 34, 87; description of processes, 83-85; output, 85; workers in, 121-122; attempts to suppress in colonies, 153; cost of, 170; number of, 182, 183.

Reformed Presbyterians, 37.
Rentgen, Clemens, 105, 190.
"Report on Manufactures," 164.
Restrictions on colonial ironworks, 151, 153, 154-155, 158.
Revolution, ironworks during, 41; settlers west of mountains during, 62; mining during, 70; Hessian prisoners of war, 112; ironmasters in, 137-139; difficulties of ironmasters during, 139-140; prices during, 177-178; growth of iron industry, 179; casting of cannon during, 180.
Richmond, coal from, 56.
Ridgely, Charles, 60, 189.
Rigby, John, 59.
Rigby, J. S., and Company, 188.
Ringwood, 110.
Roberts, Hugh, 97.
Rock Run Furnace, 187.
Rockdale Forge, 55, 189.
Rock Forge and Slitting Mill, 190.
Rockland Forges, 189.
Rokeby Slitting Mill, 55, 191.
Rolling and Slitting Mills, See Slitting Mills.
Rolls, 88, 105.
Roosevelt, Nicholas, 102.
Ross, George, 58, 136, 139.
Ross, George, and Company, 188, 189.
Ross, John, 131.
Ross, John, and Company, 188.
Rotation of crops, 35.
Rowley, 18.
Roxborough Furnace, 188.
Ruck, John, 18.
Russia, 154, 164.
Rutter's Forge, 144, 187.
Rutter, Thomas, pioneer ironmaster, 41, 43; erects Colebrookdale Furnace, 50, 170; enterprise in Delaware, 53; nationality, 131; early career, 133, 172; ironworks of, 187.
Rutter, Thomas, Jr., 134.
Rutter, Thomas, and Company, 187.

S

Sadsbury Forge, 57, 192.
Salem, 18.
Salford Forge, 133, 189.
Sally Ann Furnace, 32, 52, 136, 190, 193.
Salt pans, 180.
Sanderson, Francis, 59.
Sand molds, 78, 81-82.
San Salvador, 11.
Sarum Iron Works, 54, 174, 178, 182, 187, 188.

Savage, Samuel, Jr., 133.
Sawmills, 35, 170.
Schoepf, Dr. John D., 41, 70.
Schools, 36-37.
School teachers, 36.
Schuyler Copper Mines, 100.
Schuylkill Forge, 191.
Schuylkill River, 40, 50, 103.
Schuylkill Valley, ruins of iron-
    works in, 29; at the end of eight-
    eenth century, 36; travelers in,
    41; ironworks of, 50-53; stone
    of, 79; ironmasters of, 144; steel
    furnaces erected, 158; capital,
    171-172; ironmasters exchange
    iron for merchandise, 175.
Scituate, 17.
Screw propeller, 101.
Seaton Iron Works, 73.
Seidel, P., 192.
Seyfert Forge, See Gibraltar Forge.
Sharp and Curtenius, 100.
Shaw, John, 118.
Shearwell Furnace, 188, See also
    Oley Furnace.
Sheet iron, 88, 161, 178.
Sheffield, 92.
Shipbuilding, 24-25, 182.
Shoenberger, Dr. Peter, 61, 131, 191.
Shöffer, Jacob, 133, 188.
Shrewsbury, 20, 150.
Signboards, 39.
Silica, 13.
Silver, mined in Tasco, 12; used by
    Aztecs, 12; used by Incas, 13;
    sent to Europe, 14; sought in
    Pennsylvania, 25; coins in Penn-
    sylvania, 36, 174, 175.
Slag, 29, 80, 85, 145. See also cinder
    iron.
Slaves, See Negro workers.
Slit iron, 88, 151, 161, 164, 182.
Slitting mills, 45; first in Penn-
    sylvania, 55; description of proc-
    esses, 88, 89, 156; war prisoners
    at work on, 112-113; workers in,
    113, 117; restrictions imposed by
    Parliament on, 153, 155; cost of,
    170; erected during Revolution,
    181; development of, 182; num-
    ber of, 183.
Slocum, E. and B., 191.
Slocum's Hollow Bloomery, 191.
Slough, Matthias, 135, 137, 139,
    141.
Smeaton, John, 82, 95.
Smith, James, 59, 136, 139, 141.
Smith, Thomas, 57, 188.
Smith, W. H., 190.
Smith, William, 57, 188.
Soho Works, 102.

Solebury Forge, 54, 189.
Soundwell Forge, 191.
Southampton Adventurers, 16.
South Mountain, 59.
Spang Forge, See Oley Forge.
Spaniards, 11, 12, 13.
Speedwell Forge (Berks County),
    192.
Speedwell Forge (Lebanon Coun-
    ty), 57, 58, 188.
Spotswood, Governor Alexander,
    16, 26, 152.
Spring Creek Forge, 61, 191.
Spring Forge (No. 1), 144, 187.
Spring Forge (No. 2), 143, 188.
Spring Forge (No. 3), 59, 189.
Spring Furnace, 192.
Spring Grove Forge, 57, 134, 144,
    190.
Spring Hill Furnace, 64, 191.
Springton Forge, 54, 189.
Spruce Creek Forge, 61, 191.
Stamping mills, 85, 97.
Staves, exports of, 35.
Steam boats, 101.
Steam engine cylinders, 99.
Steam engines, 99-103.
Stedman, Alexander, 57, 141.
Stedman, Charles, 57, 141.
Steel, tools, 85; description of, 91;
    improvements of, 86-87, 103-104;
    restrictions on making, 153; en-
    couragement for making, 158-
    159, 180; duties on, 160, 161, 164;
    development of manufacture of,
    182.
Steel furnaces, for making blister
    steel, 45; in Philadelphia, 55;
    German method tried, 85; de-
    scription of processes, 86-87; at-
    tempts to suppress, 153; erection
    prohibited in colonies, 155; in
    Schuylkill Valley, 158; cost of,
    170; erected during Revolution,
    181; progress after Revolution,
    182; number of, 183.
Stevenson, George, 58, 59, 136, 188.
Stiegel, Elizabeth, 144.
Stiegel glassware, 141.
Stiegel, Henry William, at Eliza-
    beth Furnace, 37; ironmaster,
    43; rise of, 57, 141-142; nation-
    ality, 131; bankrupt, 142; "Feast
    of Roses," 142-143; cost of iron-
    works, 170.
St. James Church Bloomery, 53.
Stockholm, 154.
Stone implements, 13.
Store, description of plantation, 34;
    prices, 124-125.

Stove plates, 81, 96, 119, 120, 176, 181.
Stoves, 33, 97-99, 102, 135.
St. Petersburg, 183.
Stuyvesant, Governor, 22.
Sunderland, 103.
Supreme Executive Council, 139.
Susquehanna River, 40, 56, 61, 141.
Susquehanna Valley, 29, 56-60.
Sweden, 22, 150, 153, 154, 164.
Swedenborg, Emanuel, 40-41.
Swedes, 22.
Swope, Benedict, and Company, 188.
Swords, 11.
Sulphur, 13.
Sylvan Forge, 64, 191.

T

Tariff, 159--161, 162-165.
Taunton, 17.
Taverns, 39.
Taylor, George, at Durham, 42, 54; nationality, 131; signer of Declaration of Independence, 136; in Revolutionary army, 139; picture, 143; clerk, 172.
Taylor, John, 54, 123, 131, 136, 174, 188.
Teamsters, 118, 124.
Thirty Years' War, 22.
Thomas, Gabriel, 24, 26.
Thompson, William, 58, 139.
Thornburg Forge, 189.
Thornburg, Joseph, 139.
Thornburg, Robert, 59, 189.
Thornburg, Robert, and Company, 189.
Thornbury Forge, 190.
Thornton, William, 101.
Tilt hammer, See Plating Mills.
Tinicum Island, 22, 23.
Tinners, 182, 183.
Tin plate, 88, 89.
Tin plate works, 56, 113.
Tobacco, 14.
Toledo steel, 91.
Tonnage duties, 160.
Transportation of iron, 39-40, 61, 177.
Travelers, 40.
Treaty of Aix-la-Chapelle, 154.
Trenton, 55.
Truckenmiller, John, 141, 189.
Tucker, Tempest, 61, 189.
Tulpehocken Eisenhammer, See Charming Forge.
Tunnel-head, 78.
Turnbull, Marmie and Company, 141, 172.
Turnbull, William, 132, 171.

Turnbull, William, and Company, 190.
Turner, Daniel, 61, 191.
Turner's Iron Works, 191.
Turner, Joseph, 54, 57, 132.
Tutors, 36.
Tuyère, 78, 79.
Twaddell's Forge, 189.
Twaddell, William, 189.

U

Undree, General Daniel, 51, 70, 137-139.
Union Forge (Blue Mountains), 188.
Union Forges (Berks County), 192.
Union Furnace and Forge (Fayette County), 64, 190, 193.
Utrecht, Treaty of, 180.

V

Valley Forge, size of plantation, 31; burning of forge, 42; Washington at, 51; Negroes employed at forge, 115; forge workers at, 124; Isaac Potts at, 139; ironworks, 187, 189.
Van Leer, Samuel, 190.
Vincent Forge and Steel Furnace, 158, 188.
Vincent Slitting Mill, 52, 190.
Virginia, first ironworks, 14-16; Governor Spotswood, 26, 152; failure of crops, 62; William Byrd, 150, 152; first shipments of bar iron from, 153; English capital in, 171.
Virginia Company of London, 14, 16, 24.
Voight, Henry, 101, 103, 104, 189.

W

Wages, 45, 110, 118-125, 178.
Walker, Daniel, and Company, 187.
Wampanoag War, 19.
War of the Austrian Succession, 153.
Warwick Furnace, buildings, 33; George Whitefield at, 37; George Washington at, 42; established, 51, 187; rivals English furnaces, 51; mines, 70; consumption of charcoal, 75; output, 79; Robert Grace at, 97, 134, 145; casting of parts for steam engines, 103; Anna Nutt and Company, owners of, 111; scarcity of labor, 111; teamsters, 124; Potts and Rutter families at, 144; cost of production at, 177; casting of cannon, 180; height and size of, 193.

Washingtonburg, 181.
Washington, George, 40, 42, 51, 86, 136.
Water wheel, 34, 55, 82, 84-85, 88.
Watt, James, 99, 102.
Wayne, Anthony, 61, 62.
Webb, J., 192.
Weiss, G. M., 188.
Wells, Richard, 104.
Welsh Mountains, 56-57.
Western Pennsylvania, ores of, 69; prices of commodities, 128.
West India Company (Swedish), 23.
West Indies, 34, 40, 111, 141, 179.
Westmoreland Iron Works, 41, 64, 170, 171, 191.
Wheelwrights, 182, 183.
Whitefield, George, 37.
Whitesmiths, 90, 182, 183.
Whittemore, Amos, 104.
Williams, Roger, 19.
Wilmington, 22, 56, 104.
Wilson, James, enterprise with Mark Bird at Delaware Works, 55, 105; nationality, 131; signer of Declaration of Independence, 136; holdings in ironworks, 143; secures Dutch capital, 171.
Windsor Forge (Berks County), 189.
Windsor Forges (Lancaster County), 56, 111, 133, 136, 187.
Windsor Furnace, 136, 189.
Wines, imported, 38.
Winthrop, John, Jr., 16, 18, 19.

Wire, 89-90, 104.
Wire mills, 89-90, 182.
Wishon, Conrad, 124.
Withers, M., 192.
Women workers, 36, 115-116.
Wood, Michael, 145.
Woodcutters, 123.
Wooddrop, Alexander, 170, 171.
Wooddrop, Alexander, and Company, 187.
Woodhouse, James, 145.
Wool, 35.
Woolen merchants, 149.
Wootz Steel, 91.
Workers, English, 16; Negro, 26, 111, 113, 114, 115, 118, 124; homes of, 32-33; amusements of, 38; wages, 45, 118-125; miners, 71, 124; colliers, 74, 75, 76, 108, 118, 123-124; scarcity of, 108-109; European, 110; types of, 110; contract, 110, 118; Indian, 111, 117; and wars, 111-113; indentured, 117-118; teamsters, 118-124; wages, 118-125; blast furnace, 119-121; bloomery, 121; refinery forge, 121-123; woodcutters, 123; laborers, 124.
Wrought iron, 77, 90.

Y

York, 39.
York, Duke of, See James, Duke of York.
York, Thomas, 131, 135, 136.
Youghiogheny Forge, 64, 192.

Library of
Early American Business And Industry

I.   John Leander Bishop, A HISTORY OF AMERICAN MANUFACTURES FROM 1608 TO 1860, with an introduction by Louis M. Hacker, 3 volumes.

II.   Albert S. Bolles, THE INDUSTRIAL HISTORY OF THE UNITED STATES, Copious Illustrations, with an introduction by Louis M. Hacker.

III.   Freeman Hunt, LIVES OF AMERICAN MERCHANTS, with an introduction by Louis M. Hacker, 2 volumes.

IV.   George S. White, MEMOIR OF SAMUEL SLATER, Illustrated with engraving, woodcuts and folding diagram.

V.   Rolla M. Tryon, HOUSEHOLD MANUFACTURES IN THE UNITED STATES, 1640-1860. A study in Industrial History.

VI.   J. D. B. DeBow, THE INDUSTRIAL RESOURCES, etc. of the Southern and Western States, 3 volumes.

VII.   TENCH COXE, A VIEW OF THE UNITED STATES OF AMERICA, with folding tables.

VIII.   Charles F. Adams, Jr., and Henry Adams, CHAPTERS OF ERIE and other Essays.

IX.   Stuart Daggett, RAILROAD REORGANIZATION.

X.   Stuart Daggett, HISTORY OF THE SOUTHERN PACIFIC.

XI.   Nelson Trottman, HISTORY OF THE UNION PACIFIC, a financial and economic survey.

XII.   Howard D. Dozier, A HISTORY OF THE ATLANTIC COAST LINE RAILROAD.

XIII. Timothy Pitkin, A STATISTICAL VIEW OF THE COM-
MERCE OF THE UNITED STATES OF AMERICA.

XIV. Katherine Coman, ECONOMIC BEGINNINGS OF THE
FAR WEST, 2 volumes.

XV. William R. Bagnall, THE TEXTILE INDUSTRIES OF
THE UNITED STATES.

XVI. Witt Bowden, THE INDUSTRIAL HISTORY OF THE
UNITED STATES.

XVII. Melvin T. Copeland, THE COTTON MANUFACTURING
INDUSTRY OF THE UNITED STATES.

XVIII. Blanche E. Hazard, THE ORGANIZATION OF THE BOOT
AND SHOE INDUSTRY IN MASSACHUSETTS BE-
FORE 1875.

XIX. Albert Gallatin, REPORT OF THE SECRETARY OF THE
TREASURY ON THE SUBJECT OF ROADS AND
CANALS, 1807.

XX. Henry S. Tanner, A DESCRIPTION OF THE CANALS
AND RAILROADS OF THE UNITED STATES.

XXI. J. Warren Stehman, THE FINANCIAL HISTORY OF
THE AMERICAN TELEPHONE AND TELEGRAPH
COMPANY.

XXII. Kathleen Bruce, VIRGINIA IRON MANUFACTURE IN
THE SLAVE ERA.

XXIII. Abraham Gesner, A PRACTICAL TREATISE ON COAL,
PETROLEUM AND OTHER DISTILLED OILS, revised
and enlarged by George W. Gesner.

XXIV. Alexander Hamilton, INDUSTRIAL AND COMMERCIAL
CORRESPONDENCE OF ALEXANDER HAMILTON
ANTICIPATING HIS REPORT ON MANUFACTURES,
edited by Arthur H. Cole. With a Preface by Prof. Edwin
F. Gay.

XXV. Lewis Henry Haney, A CONGRESSIONAL HISTORY OF RAILWAYS IN THE UNITED STATES, 2 volumes in one.

XXVI. Adam Seybert, STATISTICAL ANNALS. Quarto.

XXVII. Samuel Batchelder, INTRODUCTION AND EARLY PROGRESS OF THE COTTON MANUFACTURE IN THE UNITED STATES.

XXVIII. Tench Coxe, A STATEMENT OF THE ARTS AND MANUFACTURES OF THE UNITED STATES OF AMERICA FOR THE YEAR 1810.

XXIX. (Louis McLane), DOCUMENTS RELATIVE TO THE MANUFACTURES IN THE UNITED STATES (Executive Document No. 308, 1st Session, 22nd Congress.

XXX. B. F. French, THE HISTORY OF THE RISE AND PROGRESS OF THE IRON TRADE OF THE UNITED STATES.

XXXI. Frederick L. Hoffman, HISTORY OF THE PRUDENTIAL INSURANCE COMPANY OF AMERICA, 1875-1900.

XXXII. Charles B. Kuhlman, DEVELOPMENT OF THE FLOUR MILLING INDUSTRY IN THE UNITED STATES.

XXXIII. James Montgomery, A PRACTICAL DETAIL OF THE COTTON MANUFACTURE OF THE UNITED STATES OF AMERICA.

XXXIV. Henry Varnum Poor, HISTORY OF THE RAILROADS AND CANALS OF THE UNITED STATES.

XXXV. Henry Kirke White, HISTORY OF THE UNION PACIFIC RAILWAY.

XXXVI. Frank B. Copley, FREDERICK W. TAYLOR, FATHER OF SCIENTIFIC MANAGEMENT, 2 volumes.

XXXVII. Edward Winslow Martin, HISTORY OF THE GRANGE MOVEMENT: or, The Farmer's War Against Monopolies.

XXXVIII. J. T. Henry, THE EARLY AND LATER HISTORY OF PETROLEUM.

XXXIX   Arthur C. Bining, PENNSYLVANIA IRON MANUFAC-
TURE IN THE EIGHTEENTH CENTURY.

XL   Thomas F. De Voe, THE MARKET BOOK, a History of
the Public Markets in the City of New York.

XLI   James M. Swank, HISTORY OF THE MANUFACTURE
OF IRON IN ALL AGES and particularly in the United
States from Colonial Times to 1891. 2nd Edition (1892).

XLII   Charles H. Ambler, A HISTORY OF TRANSPORTATION
IN THE OHIO VALLEY; with Special Reference to its
Waterways, Trade and Commerce from the Earliest Period
to the Present Time (1932).

XLIII   Pery W. Bidwell and John I. Falconer, HISTORY OF
AGRICULTURE IN THE NORTHERN UNITED STATES
1620-1680 (1925).

XLIV   Jules I. Bogen, THE ANTHRACITE RAILROADS. A Study
in American RAILROAD ENTERPRISE.

XLV   Helen Cowan, CHARLES WILLIAMSON; Genesee Promoter, Friend of Anglo-American Rapprochment (1941).

XLVI   J.B.D. DeBow, STATISTICAL VIEW OF THE UNITED
STATES . . . Being a Compendium of the Seventh Census
to which are Added the Results of Every Previous Census
. . . in Comparative Tables (1854).

XLVII   Lewis C. Gray, HISTORY OF AGRICULTURE IN THE
SOUTHERN UNITED STATES TO 1860. 2 vols. (1933).

XLVIII   Robert Henriques, BEARSTED. A Biography of Marcus
Samuel, First Viscount Bearsted and Founder of "Shell"
Transport and Trading Company (1960).

XLIX   Malcolm Maclaren, THE RISE OF THE ELECTRICAL
INDUSTRY DURING THE NINETEENTH CENTURY
(1943).

L   Lorenzo Sabine, REPORT ON THE PRINCIPAL FISH-
ERIES OF THE AMERICAN SEAS (1852).

LI   Lawrence H. Seltzer, THE FINANCIAL HISTORY OF
      THE AMERICAN AUTOMOBILE INDUSTRY. A Study
      of the Ways in Which the Leading American Producers of
      Automobiles Have Met Their Capital Requirements (1928).

LII   George W. Stocking, THE OIL INDUSTRY AND THE
      COMPETETIVE SYSTEM. A Study in Waste (1925).

LIII William Strickland, et al, eds., REPORTS, SPECIFICA-
      TIONS AND ESTIMATES OF THE PUBLIC WORKS
      IN THE UNITED STATES OF AMERICA (1841).

LIV   C. W. Ackerman, GEORGE EASTMAN. With an Introduc-
      tion by E. R. A. Seligman (1930).

LV   James Hall, STATISTICS OF THE WEST at the close of
      the year 1836.

LVI   Samuel Hazard, Ed., HAZARD'S UNITED STATES COM-
      MERCIAL AND STATISTICAL REGISTER. Documents,
      Facts, and other Useful Information Illustrative of the His-
      tory and Resources of the American Union, and of Each
      State, 6 volumes.

LVII Freeman Hunt, WORTH AND WEALTH, a Collection of
      Maxims, Morals and Miscellanies for Merchants and Men
      of Business.

LVIII Samuel A. Mitchell, MITCHELL'S COMPENDIUM OF
      THE INTERNAL IMPROVEMENTS OF THE UNITED
      STATES, Comprising Notices of all the Most Important
      Canals and Railroads.

LIX   E. D. Kennedy, THE AUTOMOBILE INDUSTRY, 1941.